ATMOSPHERIC OPTICS

Edited by Nikolai B. Divari

Chairman, Department of Higher Mathematics
Odessa Polytechnic Institute

Translated from Russian by
Stephen B. Dresner

 CONSULTANTS BUREAU · NEW YORK · 1970

The original Russian text, published by Nauka Press in Moscow in 1968 and corrected by the editor for this edition, constitutes the proceedings of conferences on atmospheric optics sponsored by the committee on the optical instability of the earth's atmosphere of the Astronomical Council of the Academy of Sciences of the USSR held in Pulkovo in November and December 1965 and December 1966. The present translation is published under an agreement with Mezhdunarodnaya Kniga, the Soviet book export agency.

Н. Б. ДИВАРИ
АТМОСФЕРНАЯ ОПТИКА

Library of Congress Catalog Card Number 69-18138

ISBN-13: 978-1-4684-7881-5   e-ISBN-13: 978-1-4684-7879-2
DOI: 10.1007/978-1-4684-7879-2

© 1970 Consultants Bureau, New York
Softcover reprint of the hardcover 1st edition 1970

A Division of Plenum Publishing Corporation
227 West 17th Street, New York, N. Y. 10011

United Kingdom edition published by Consultants Bureau, London
A Division of Plenum Publishing Company, Ltd.
Donington House, 30 Norfolk Street, London, W.C. 2, England

CONTENTS

CONTENTS

SOME STATISTICAL DETERMINATIONS OF THE INSTABILITY OF TELEVISION IMAGES OF STARS

N. F. Kuprevich and A. Kh. Kurmaeva

Experimental work on a television compensator for interference from atmospheric turbulence in astronomical observations has been started at the Main Astronomical Observatory of the Academy of Sciences of the USSR. Some possibilities for using photoelectric and television methods to combat noise have already been considered [1, 2]. To determine certain fundamental conditions for the design of a television noise compensator, however, it was necessary to study and statistically process large-scale (the telescope had an optical focus of 125 m) star images, which were filmed from a kinescope screen at 25 frames/sec. It was also necessary to create, if only rather approximately, a method of determining a quality criterion for an astronomical (in this case, a point) image and, thereby, to be able to compare quantitatively the changes in the structure of the image caused by scintillation.

It should be pointed out that in conventional photoelectric recording of star-image instability, only variations in luminous flux, due to atmospheric noise, as functions of observation time are recorded. This phenomenon is characterized by the simple bivariate intensity−time system. But when one processes photographs of large-scale images of scintillations, one encounters multivariate characteristics, which reflect the variations with respect to time as well as in space (in different directions). In this case, the nature of the image actually varies not only with respect to time but is also deformed in various planes. The luminosity is redistributed in individual parts of the turbulent disk of the star image (bunches and gaps in luminosity are formed). And there are shifts and surges of parts of the image in different directions, which distort its shape. This makes it rather complicated to characterize and represent quantitatively the multivariate change in the characteristics of the turbulent image.

"Discontinuity and variability criteria" have been developed as an elementary, approximate method for comparing nonuniformities in star images. They will be discussed in greater detail below.

Oscillograms of photoelectric photometer currents served as the initial information for the processing and statistical study of the photographs. The television photographs of the stars were projected onto the input of a scanning photometer designed by N. F. Kuprevich [3] in a scale that provided 20-line scanning of the star images. This gave approximately identical resolution for transmission through the television system and through the scanning photometer. The photographs were not given complete photometric processing, since the problem included determination not of the exact variations of luminous-flux intensities in the star image but only determination of the differential and its spatial position in the turbulent disk.

Statistical Determination of Nonuniformities in Television Images of Stars

Fifteen kinescope frames of α Tauri were used for the initial statistical studies. The photographs were made with a synchronous motion picture camera at 25 frames/sec. The exposure time of each frame was 1/50 sec, and the remaining 1/50 sec was for movement of the film in the camera. Alternate-line scanning in the television equipment with 600-line definition at 25 frames/sec provided photographing of only one half-frame and a vertical definition of 300 lines. The 125-m optical focus of a reflecting telescope with an input aperture of 285 mm [4] was used to construct the optical image of the star at the input of the television system.

The oscillograms of the photocurrents of 20-line scanning of the star image, which were obtained from the scanning photometer, were measured on a P-10 enlarger; 121 points (11 points in each of 11 lines) were measured on an oscillogram. The ordinates of the oscillograms, which characterized the degree of light transmission in the film positive (which contact printed from the negative), were measured. The maximum ordinate value was set equal to unity.

As a first approximation, calculations were made to obtain a qualitative picture of the degree of variability of the shape of the star images. The number of points with low and high values of luminous-flux intensity were calculated and their ratios examined for 15 images. Analysis showed that these ratios could characterize the nonuniformity of star images. When the ratio of the number of low and high values of luminous-flux intensity is close to unity, the star image has a regular (circular) shape and a uniform distribution of luminous-flux density. Higher values (> 4) of this ratio indicate great deformation of the image.

In the next stage of operations, it was necessary to obtain quantitative characteristics of the degree of nonuniformity of the television images. The distribution of illumination over the disk of the image could be seen by Bessel or Fourier analysis [5–7]. But this is a very time-consuming process, so we used the following statistical method. Given m = 11 lines and n = 11 points along each line, the coordinates a and b of the centroids of the images are

$$
\begin{aligned}
a_j &= \left[\sum_{k=1}^{n} \sum_{i=1}^{m} i I_j(i, k) \right] : \left[\sum_{k=1}^{n} \sum_{i=1}^{m} I_j(i, k) \right]; \\
b_j &= \left[\sum_{k=1}^{n} \sum_{i=1}^{m} k I_j(i, k) \right] : \left[\sum_{k=1}^{n} \sum_{i=1}^{m} I_j(i, k) \right].
\end{aligned}
\tag{1}
$$

where i is the line number; k the point number on a line; j the image number; p the number of images (five, in our case); and $I_j(i, k)$ is the luminous-flux intensity at the point with coordinates i, k on the j-th image. Below, we shall let i = 0 and k = 0 at point (a_j, b_j). The luminous-flux intensity is calculated at each point on the average image by the formula

$$
I(i, k) = \frac{1}{p} \sum_j I_j(i, k).
$$

Then, from rather simple relationships it is possible to determine the variability criterion A for a given series of images and the discontinuity criterion D for each image in two mutually perpendicular directions

$$
A = \left\{ \sum_j \sum_k \sum_i [I_j(i, k) - I(i, k)]^2 \right\} : p \sum_k \sum_i I^2(i, k).
\tag{2}
$$

Obviously, the greater the variations in the intensity distribution over the disk of the image from star to star, the greater the absolute values of A.

We let $i = 0$ and $k = 0$ for calculation of the discontinuity criteria at points (a_j, b_j) and for the average image (a, b). Then the discontinuity criteria are determined along the lines

$$D_j' = \sum_i \sum_{\substack{k \geqslant 0 \\ k < 0}} \{[I_j(i, k \pm 1) - I_j(i, k)] - [I(i, k \pm 1) - I(i, k)]\} H(i, k) \tag{3}$$

and along those perpendicular to them

$$D_j = \sum_{\substack{i \geqslant 0 \\ i < 0}} \sum_k \{[I_j(i \pm 1, k) - I_j(i, k)] - [I(i \pm 1, k) - I(i, k)]\} H(i, k). \tag{4}$$

The discontinuity criterion for an entire image is the sum
$$D = D' + D''.$$

In expressions (3) and (4), the function H is equal to zero if the difference within the braces is less than 0, since this means that the distribution of luminous-flux intensity over the disk of a given image is normal, i.e., decreases from the center to the edges. In this case, the discontinuity criterion is also equal to 0. But when the difference within the braces is greater than 0, the distribution is not normal. Then the function H is equal to 1 and the discontinuity criterion is equal to some positive number.

For our small series of five star images, the variability criterion $A = 0.26$ and the discontinuity criterion for image
$$D_1 = 0, \quad D_2 = 2.13, \quad D_3 = 10.24, \quad D_4 = 8.39, \quad \text{and} \quad D_5 = 0.$$

Note that the discontinuity criteria are calculated with respect to an average image. Therefore, as the number of images is increased, the average approaches the ideal image, which has a uniform distribution of luminous-flux intensity over the disk with a slight decrease from the center to the edge, and its shape is almost circular. In this case, the values of the discontinuity criteria of each star in a series should be close to absolute.

At first glance, the shapes of our star imates and the corresponding values of D did not agree (Fig. 1). For example, image 3 has the more regular shape and a uniform distribution of luminous-flux intensity over the disk, but it also has the higher discontinuity criterion $D = 10.24$. Conversely, greatly deformed images 1 and 5 have $D = 0$. This happened because the average of these five images came closer to the second, fifth, and even first images than did the third. Hence, the order of increasing D from the first and fifth, successively through the second, fourth, and third images given above.

Since the variations in luminous flux in one image to a certain extent predetermine these variations at the next instant, it is interesting to calculate the correlation functions

$$B(l) = \sum_k \sum_i B(i, k, l) = \sum_k \sum_i \sum_j \{[I_j(i, k) - I(i, k)][I_{j+l}(i, k) - I(i, k)]\}, \tag{5}$$

where $l = 0, 1, 2, \ldots, p$.

Since the element of chance is present in the phenomenon under consideration, a fairly large series of observations must be made to obtain a realistic result.

It should be noted that 20-line scanning makes it possible to calculate the correlation functions for a given (i, k) point

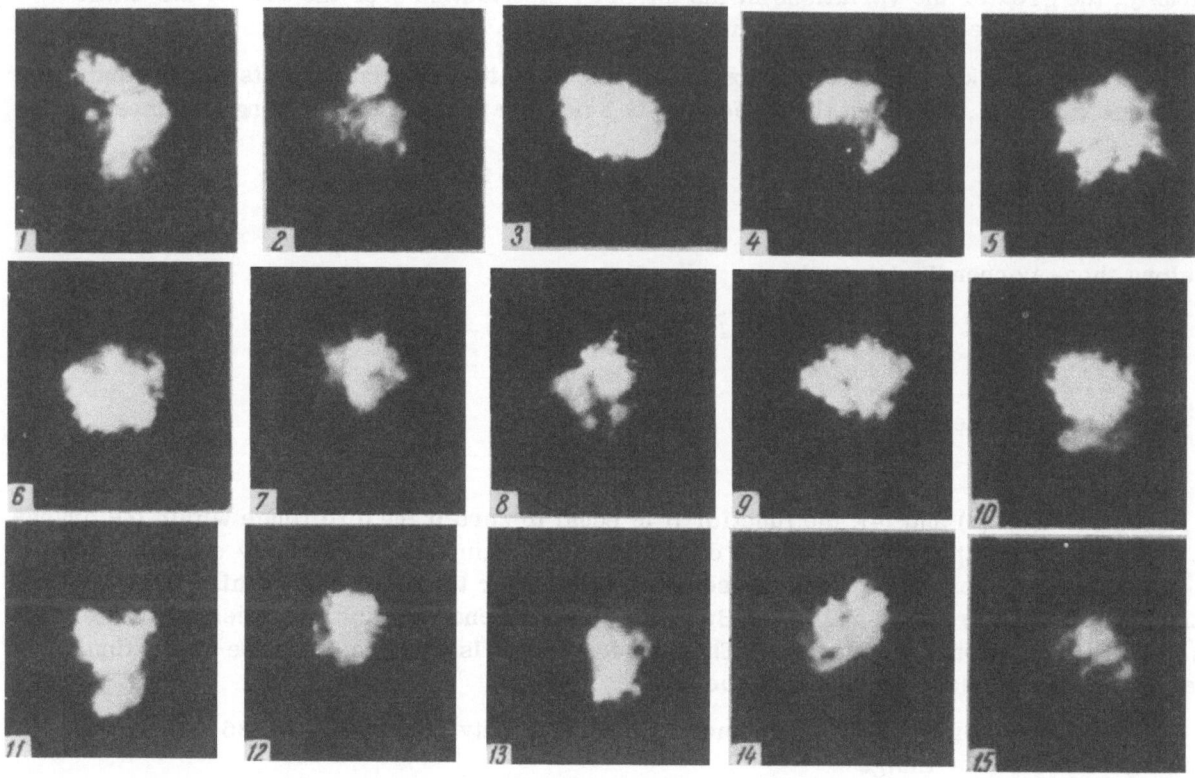

Fig. 1

$$B(i, k, l) = \sum_{j} \{[I_j(i, k) - I(i, k)] \, [I_{j+l}(i, k) - I(i, k)]\}, \tag{6}$$

and also the correlation along the lines and perpendicular to them. This is of particular interest, because it provides still another characteristic of the degree of nonuniformity of star images.

Additional calculations have shown that positive and negative images give equivalent results, which is important for reducing the processing time. When using the method described above, the time for processing the material can be reduced considerably by using an electronic computer, which we are now doing for a great number of stars.

We thank Professor T. A. Agekyan, of Leningrad State University, for valuable advice.

LITERATURE CITED

1. N. F. Kuprevich, Izv. GAO Akad. Nauk SSSR, No. 175 (1964).
2. N. F. Kuprevich, Author's Certificate, Byull. Izobret., No. 23 (1959).
3. N. F. Kuprevich and A. Kh. Kurmaeva, Izv. GAO Akad. Nauk SSSR, No. 180 (1966).
4. N. F. Kuprevich, Izv. GAO Akad. Nauk SSSR, No. 163 (1960).
5. Z. M. Kanevskii and M. I. Finkel'shtein, Fluctuation Noise and the Detection of Radio Pulses [in Russian], Gosénergoizdat, Moscow (1963).
6. A. Mareshal and M. Franson, Structure of the Optical Image [Russian translation], Izd. Mir, Moscow (1964).
7. "Image transmission carrier and image spectrum in generalized photographic system by Kinoshita," Nippon Hoso Kyokai (Japan Broadcasting Corporation), Technical Research Laboratories, Tokyo, Techn. Monogr., No. 3 (1964).

IMAGES OF THE LUNAR LIMB AND HETEROGENEITY LAYERS IN THE TROPOSPHERE

A. N. Demidova

In an earlier paper [1], we considered, among other things, the displacement of large sections of images of the solar limb and the height of the atmospheric layers that cause this phenomenon. Since the vibrations were estimated visually with the eyepiece grid, we dealt with near-maximum amplitudes. It was shown that the upper limit of atmospheric heterogeneities that were capable of causing displacement as a unit of large sections of an image of the solar limb (with a chord of about 9') was at an altitude on the order of 100 m.

Layers that effect another phenomenon on a sun image — shifting wavy deformations of the limb — are situated considerably higher, on the average, in the atmosphere [1, 2]. According to the data in [1], these heterogeneities are, as a rule, in layers at heights of from a few hundreds of meters to 2-3 km. On the average, they are at 1.5 km, i.e., much higher than the surface layers that displace large sections of sun images.

A similar phenomenon can be observed at night on the lunar limb. Such observations have been made at Pulkovo since 1961 [3]. Even the first results showed that heterogeneities that affect the edge image of an extended source are, on the average, in higher-altitude layers at night than in the daytime.

Just as with the sun, deformations are not always observed on lunar-limb images. They were entirely absent in approximately 15% of our observations. The deformation pattern sometimes has a complex structure, due to the presence of several optically active heterogeneity layers. In the period from 1961 through 1964, there were 59 observations of the lunar limb in cloudless weather with a simple deformation pattern, which indicate the presence of a single predominant perturbing layer. In these cases, the observation times were close to the standard times for pilot-balloon and radio sounding of the atmosphere at the weather stations nearest Pulkovo. The time difference does not exceed 1 h. This fact is very important, since it makes it possible to determine the height of the heterogeneity layers by comparing their direction and velocity, as calculated from observation data, with meteorological-sounding data on wind velocity and direction.

Figure 1 shows bar graphs of the height distribution of heterogeneities for the four years 1961-1964 individually and a general bar graph for the entire 59 observations. Figure 2 shows a graph of the altitude distribution by month. The heights were determined by the method described in [3]. It follows from these data that heterogeneity layers that affect images of the limb of an extended celestial source at night are, when the sky is clear, at heights of 1-8 km and mainly at 2-3 km. The mean altitude is 3 km. Seasonal variations in the height of the heterogeneity layers have not been observed.

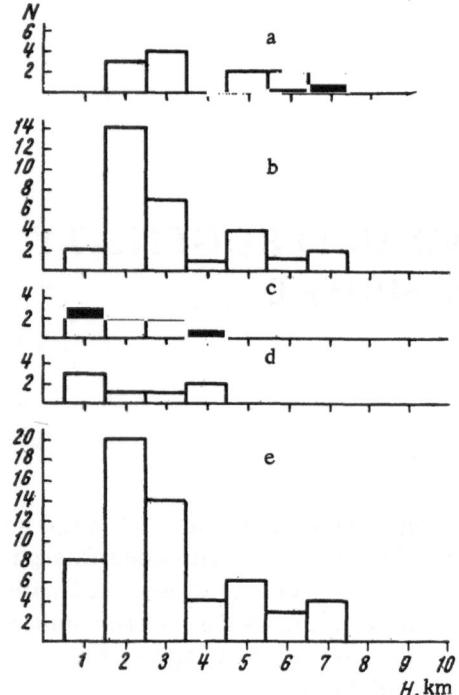

Fig. 1. Bar graphs of heterogeneity height distribution: a) 1961, b) 1962, c) 1963, d) 1964, e) 1961-1964.

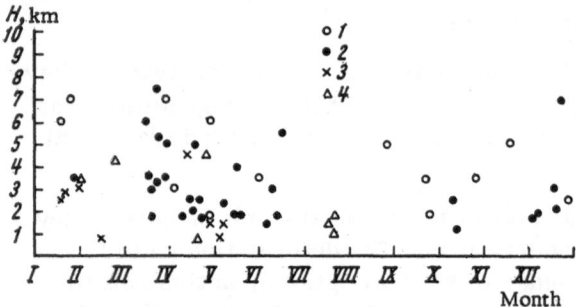

Fig. 2. Height distribution by month: 1) 1961, 2) 1962, 3) 1963, 4) 1964.

The fortunate fact that two weather stations with pilot-balloon and radio sounding are located near Pulkovo makes it possible not only to determine the height of heterogeneity layers that affect lunar-limb images but also to analyze the weather conditions, in order to show, if only in general outline, the characteristics to which formation of the heterogeneities may be related.

Analysis of aerological-sounding data for times close to the moon observation times has shown that in 42 cases out of the 59 the height of the heterogeneity coincided with an atmospheric layer, or its base, in which there was either an abrupt wind shift or a relatively large wind gradient or both. Such a situation has occurred at 0.7-7 km, but mainly at 2-3 km.

A temperature inversion or isothermy and, in a number of cases, an abrupt change in humidity have very often been observed in such a layer or in an adjacent layer. The vertical wind gradient in these layers has, in some cases, had values such as 7 m/sec at 270 m and 9 m/sec at 480 m. The wind shift has reached 100° per 1 km of height and more.

The 42 cases included several in which during the moon-observation period the wind velocity from the earth's surface to a certain level was equal to zero or was very low, i.e., about 1-2 m/sec. In these cases, the height at which heterogeneities were observed coincided with a level at which wind had just begun to be generated or had been appreciably intensified.

Twice, heterogeneities were recorded in a region of layers with a high wind velocity relative to the surrounding space. These layers were at a height of 5-7 km and at a distance of about 2 km from the lower boundary of the tropopause.

Another group consists of four cases in which the height of the heterogeneities correspond to the lower boundary of the tropopause, at 6-8 km. Also characteristic of these cases was the fact that below the heterogeneity layer there were no sudden wind shifts but there were layers with isothermy.

In 11 of the 59 cases, the heterogeneities were located considerably below the tropopause, at 1-6 km. According to sounding data, at the level of the heterogeneities at these times there were no abrupt changes in the wind or other meteorological characteristics, and only occasionally did the heterogeneity layer coincide with an inversion or isothermy.

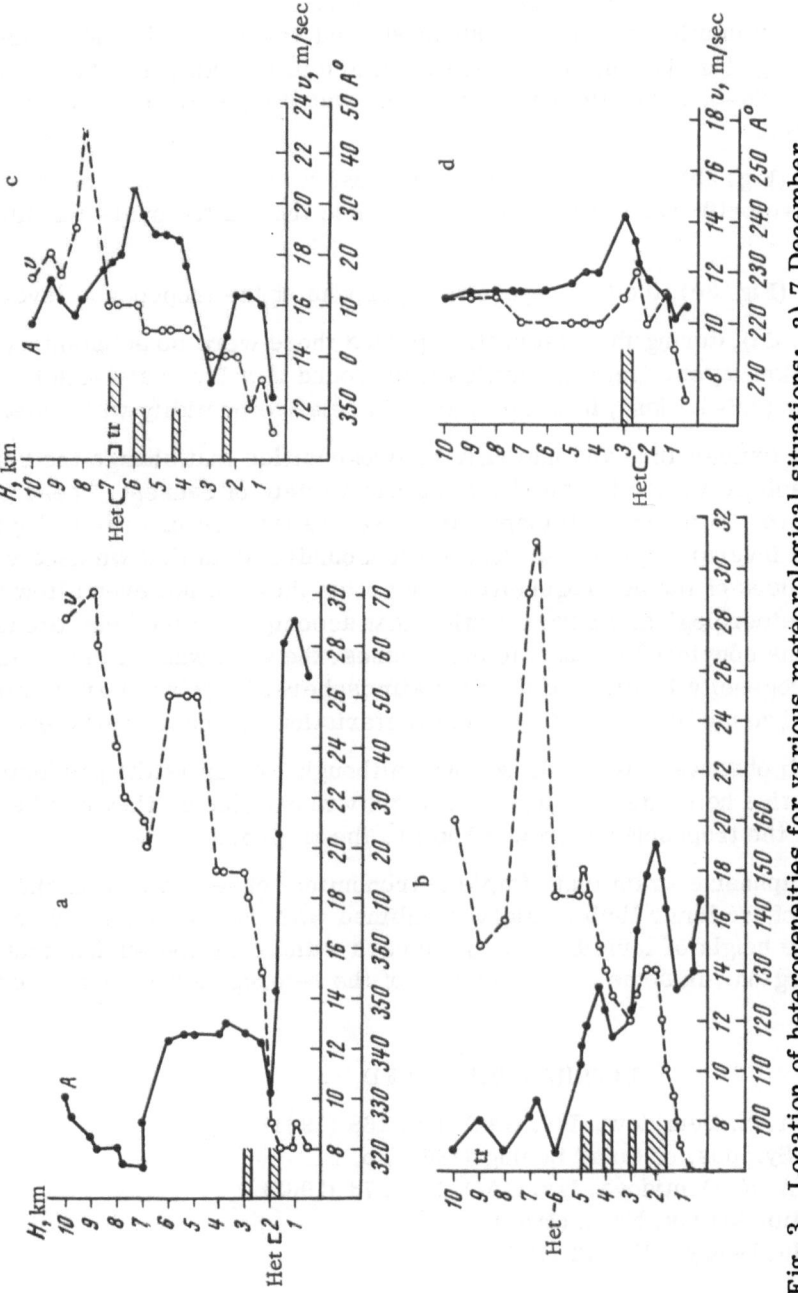

Fig. 3. Location of heterogeneities for various meteorological situations: a) 7 December 1962, b) 21 January 1961, c) 24 January 1961, d) 6 May 1962. All times: 2100 h.

As an example, Fig. 3 shows some of these situations. Altitude is plotted on the vertical axis and wind velocity (broken line) and direction (solid line) are plotted on the horizontal axis. Inversion layers (and isothermal layers) are indicated by shaded bars, and "tr" denotes the lower limit of the tropopause. The height of heterogeneities as determined from lunar-limb observations is indicated on the left of the vertical axis.

On 7 December 1962 (Fig. 3a), a heterogeneity layer was found in an area of sudden wind change: at 1.5-2 km, the azimuth variation was about 90° and the wind velocity, which had remained practically unchanged in the Pulkovo region from 500 m to 2 km, then began to increase. In the range of 2.05-2.33 km, the velocity change was 6 m/sec. An isothermal layer was found at 1.68-2.05 km.

On 21 January 1961 (Fig. 3b), heterogeneities were observed at the lower boundary of a layer in which the wind velocity was greater by 14-15 m/sec than in the over- and underlying layers.

On 24 January 1961 (Fig. 3c), the heterogeneity layer was at the tropopause level.

On 6 May 1962 (Fig. 3d), during the observation period there were no substantial wind changes in the lower 10-km layer. Heterogeneities were found in a layer at 2-2.6 km that adjoined an inversion layer (2.6-3.1 km), in which a drop in relative humidity was observed.

It is well known that temperature and humidity heterogeneities that change the refractive index occur in the atmosphere very often and have a great variety of causes. These causes are primarily inversion layers. In the free atmosphere, these are friction inversions, dynamic inversions, compression inversions, etc. [4]. The radio-sounding data that we used were insufficient to show the causes of the heterogeneities, and often they did not even allow us to establish the changes in meteorological characteristics that accompanied the heterogeneity layers. But our observations convinced us that, in most cases, the wind was an important factor in the formation of heterogeneity layers. In the free atmosphere, large wind gradients at the level of contact between two media with different characteristics can cause turbulent zones.

From the lunar-limb observations it follows that, although 2-3 km is the predominant height of the optically active heterogeneity layers at night without clouds, they can be found at practically any height in the troposphere, from ~1 km to the tropopause.

It is important to emphasize again that simple astronomical observations (of the solar limb in the daytime and of the lunar limb at night) combined with radio-sounding data make it possible to determine the height of invisible moving layers with refractive-index heterogeneities in the visible region. In some cases the size of the heterogeneities can be estimated [5].

LITERATURE CITED

1. N. V. Bystrova and A. N. Demidova, Izv. GAO, No. 169 (1961).
2. M. Minnaert, Proc. Sympos. on Solar Seeing (1961), p. 21.
3. N. V. Bystrova and A. N. Demidova, Izv. GAO, No. 173 (1963).
4. V. N. Troitskii, Radiotekhnika, No. 1 (1956).
5. A. N. Demidova, Soln. Dannye, No. 12 (1964).

MEASURING LIMB VIBRATIONS OF SOLAR IMAGES

M. A. Kallistratova

In studying fluctuations of the arrival angle of light that has passed through an atmospheric layer (or, in other words, "vibrations" of star and solar- and lunar-limb images), one of the most important problems is that of establishing quantitative relationships between the statistical characteristics of these fluctuations and the weather conditions on the propagation path. But rather few attempts have been made to establish such relationships. This is evidently due to difficulties in organizing detailed studies of fluctuations of the atmospheric refractive index simultaneously with vibration observations. Moreover, the great variety of weather conditions that can exist on the beam path make it difficult to compare data obtained in different locations and at different times. The literature contains rather contradictory opinions on whether vibration is caused mainly by refractive-index fluctuations in the lowest atmospheric layer or whether an important role is played by localized high-altitude layers with a wavy refractive-index structure. Similar disagreements arise in explanations of ultralong UHF propagation. Some researchers attribute it to incoherent scattering by turbulent fluctuations; others, to reflection from local, long-lived layers.

In a paper on displacement of parts of an image of a 9' section of the solar disk, N. V. Bystrova and A. N. Demidova [3] concluded that the heterogeneities responsible for this displacement were in a surface layer extending for several hundreds of meters. This agrees with the conclusions of Stock and Kopper [4], which were based on analysis of numerous observations, and with the conclusions of I. G. Kolchinskii [5], who studied the correlation between deviations of star trails on photographic plates. But besides these displacements, approximately 50% of the observations of Demidova and Bystrova indicate wavy motions along the solar limb. Comparing the observations with pilot-balloon sounding data, they obtained, with a rather vast amount of material, a correspondence between the observed azimuth and velocity of the wave motions along the limb and the wind velocity and direction at any altitude in the troposphere [3, 6]. By measuring the direction of the wave motions on images of the lunar limb simultaneously with the direction of the moving shadows [7], Demidova and Bystrova showed that both directions were the same. This indicates that the moving shadows and vibrations were caused by the same layer of atmospheric heterogeneities. According to [8], the altitudes of the perturbing layers are from 0.3 to 2.5 km in the daytime and from 1.5 to 9.0 km at night. Unfortunately, such observations do not allow the vibration intensity to be related to any of the parameters of the layer, nor do they permit estimation of the relationship between the energy of displacement of large image sections and the energy of wave motions along the limb.

Dommanget [9] and Zinchenko [10] have compared the vibration intensity and quality of star images with the vertical wind, temperature, and humidity profiles. Zinchenko showed that tropospheric layers with unstable stratification always reduce image quality. Dommanget, who analyzed 60 nights' observations, concluded that there was no correspondence between any feature of the mean-temperature profile and vibration intensity, but he did discover a relationship between vibration and discontinuities in the vertical humidity profile. He does not provide a convincing explanation for this, inasmuch as he himself cites data indicating that the refractive index for the visible region is considerably more dependent upon air temperature than humidity.

Analysis of the conclusions of the various authors indicates that image vibration can be caused by refractive-index fluctuations in the lower, usually more turbulent part of the atmosphere as well as in higher layers. The effect of the upper layers can predominate or be entirely absent, depending upon weather conditions.

We decided to make quantitative comparisons of the variance and frequency spectrum of the vibrations of solar-limb images with weather conditions and to find out whether it is possible to calculate the vibrations when weather conditions are well-known.

Measurement Conditions and Apparatus

The main measurements were made during July and August of 1963 and 1964 in the open steppe near Tsimlyansk on level ground in anticyclonic hot weather. The measurements were made on clear days and on days with small amounts of cumulus clouds. Well-developed convection in the surface layer was typical of these days.

The solar-disk images were obtained using an ATsU-24 horizontal mounting whose coelostat was driven by a synchronous motor, which was powered by a quartz-crystal oscillator. The mounting was placed on a special foundation 3 m from ground level. The limb displacements were recorded with the electronic tracking system described in [1]. A slight change was made in the system [1] for the 1964 measurements: a right-angle prism was placed near the focal point of the telescope, which made it possible to aim any part of the solar limb at the slit. Particular attention was given to reducing the ray path within the telescope itself, to ensuring good intermixing of the air within the telescope, to the smoothness of motion of the coelostat, and to eliminating mirror vibration. An AZT-7 Maksutov meniscus telescope with a focal length of 2 m was used as the objective. It was placed immediately behind the coelostat, so that the ray path between the mirrors and the objective did not exceed 2 m. The periodic drive error of the coelostat was almost one-fifth of that of the coelostat of the Main Astronomical Observatory (Pulkovo), with which the measurements were made in 1962 [1]. For a period of 8 min (the passage time for one tooth of the main worm gear), the peak-to-peak vibration of an image of the vertical solar limb was not over 5". These vibrations were filtered out by RC networks in the measuring circuit. The periodic drive error of the coelostat did not affect the position of the horizontal limb image. Vibrations of the telescope foundation, and also of the coelostat and the auxiliary mirror, in strong wind were studied with seismic detectors. With wind velocities of ≤ 6 m/sec, vibrations of the mirrors caused image displacements of not over 0".2. With higher wind velocities at the telescope level, the amplitude of the mirror vibrations increased greatly. These vibrations were sharply resonant and easily detected in the frequency spectrum of the signal.

The input aperture of the tracking-system instrument was placed in the focal plane of the AZT-7 objective. The additional magnification ($\times 5$) within the instrument gave the entire system an equivalent focus of 10 m. The input aperture of the telescope was determined by the mirror aperture of the tracking-system and was $d = 50$ mm.

The size of the limb-image section whose mean displacement was recorded in 1963 was 20". In 1964, the section sizes varied from 6" to 4'. The output voltage of the tracking system was proportional to the limb displacement relative to the central position. The variance of the arrival-angle functuations $\sigma_\varphi^2 = \overline{\Delta\varphi^2}$ was measured with a square-law function generator, which was developed at the Institute of Atmospheric Physics, whose scale was calibrated in square seconds. A 25-channel low-frequency random-process analyzer, which was also developed at the Institute, was used for frequency analysis.

Characteristics of Turbulent Refractive-Index Fluctuations

The most convenient characteristic of the intensity of turbulent temperature fluctuations in the real atmosphere is the structure constant C_T, which is defined by the Kolmogorov–Obukhov "$2/3$ law:"

$$\overline{[T(r)-T(r+\rho)]^2} = C_T^2\rho^{2/3}. \tag{1}$$

For the visible region, refractive-index fluctuations are chiefly determined by temperature fluctuations, and the structure constants for refractive-index fluctuations C_n and temperature fluctuations C_T are related by [1]

$$C_n = \frac{80\cdot10^{-6}p}{T^2}C_T. \tag{2}$$

In the surface layer, C_T is defined in terms of the vertical profiles of mean wind velocity and temperature [2]

$$C_T = \frac{0.54}{z^{1/3}}\frac{dT(z)}{d\ln z}A(\text{Ri}), \tag{3}$$

where $A(\text{Ri})$ is a known function of the Richardson number, which characterizes the turbulence instability [12].

For the conditions of well-developed turbulence under which the vibration measurements were made, the dependence of C_T upon altitude is known [13, 14]:

$$C_T^2(z) = C_T^2(z_1)\left(\frac{z}{z_1}\right)^{-4/3}. \tag{4}$$

Selective aircraft sounding has shown that, on the average, formula (4) was fairly well justified under the measurement conditions. Thus, under the experimental conditions, if we measure the gradients of mean wind velocity and temperature in the surface layer, we can calculate C_n on the light-ray path fairly reliably by formulas (2), (3), and (4).

Vibration Variance as a Function of C_n and the Form of the Frequency Spectrum for Point and Extended Sources

The statistical theory of wave propagation in a turbulent medium developed by V. I. Tatarskii gives the following expression for arrival-angle fluctuations of a plane wave from a point source:

$$\sigma_\varphi^2 = 2.8\,d^{1/3}\sec\theta\int_{z_1}^{\infty}C_n^2(z)\,dz, \tag{5}$$

where θ is the zenith angle of the source and z_1 is the height above the earth's surface of the receiving objective, which has diameter d.

_____

* The numerical coefficient in this formula had an error in [2, 15].

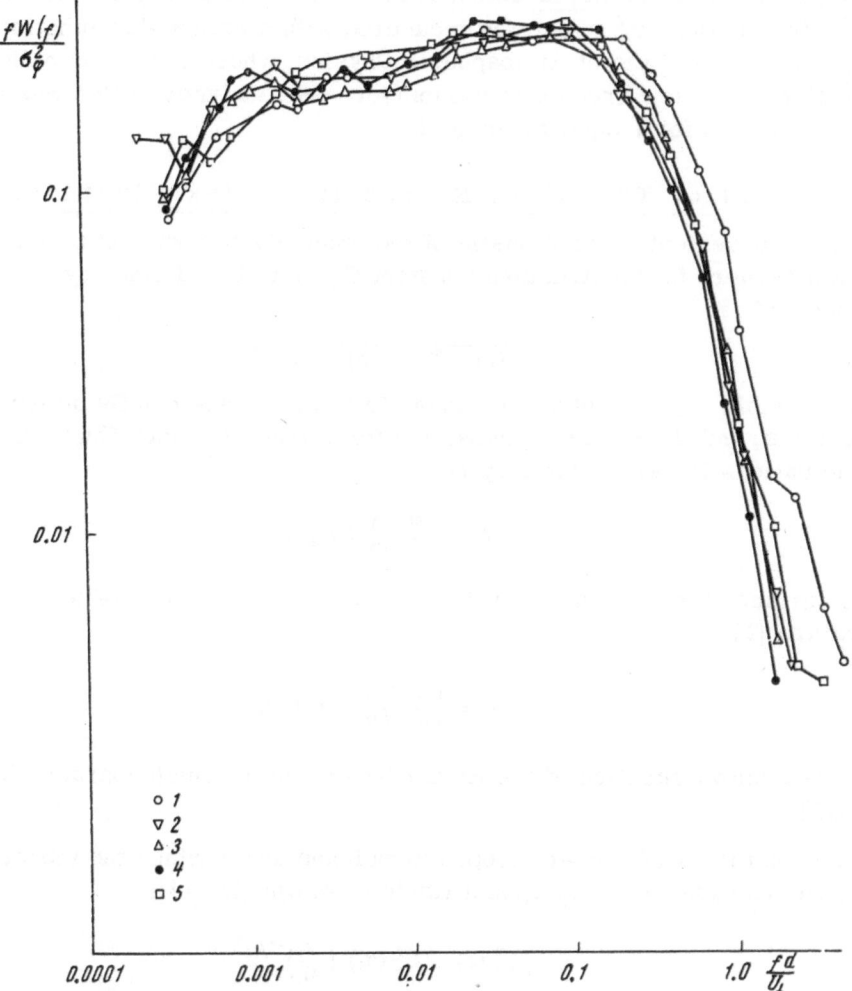

Fig. 1. Measured vibration spectra of solar-limb image for: 1) $\gamma = 9''$, 2) $\gamma = 36''$, 3) $\gamma = 72''$, 4) $\gamma = 144''$, 5) $\gamma = 280''$.

If we take formula (4) into account and integrate, we obtain from (5)

$$\frac{\sigma_\varphi^2}{\sec \theta} = 2.8 \, d^{1/s} \cdot 3 z_1 C_n^2(z_1). \tag{6}$$

It follows from formula (6) that under conditions when (4) is valid the integral effect of the entire atmosphere is equivalent to the effect of a layer whose structure constant is unchanged, $C_n = C_n(z_1)$, and whose thickness is three times z_1. In spite of this, however, it is difficult to reduce significantly fluctuations of the arrival angle of rays that have passed through the entire atmosphere by raising the receiving objective above ground on level terrain on foundations, towers, etc. It follows from (4), (2), and (6), that, as the objective is elevated, the root-mean-square fluctuations of the arrival angle must decrease as $z^{-1/6}$, i.e., in order to reduce the vibration by one-half, the height of the objective must be increased by a factor of 64.

The energy frequency spectrum of the fluctuations is defined as the Fourier transform of the autocorrelation function

$$W(f) = \frac{1}{2\pi} \int_{-\infty}^{\infty} B(\tau) e^{i\omega\tau} d\tau. \tag{7}$$

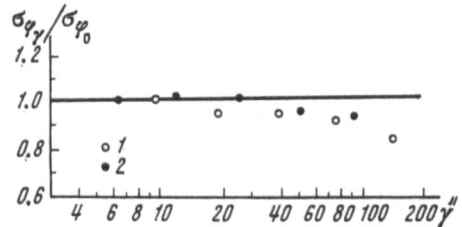

Fig. 2. Measured ratios of vibration variances of point and extended sources: 1) August 1964, 2) July 1964.

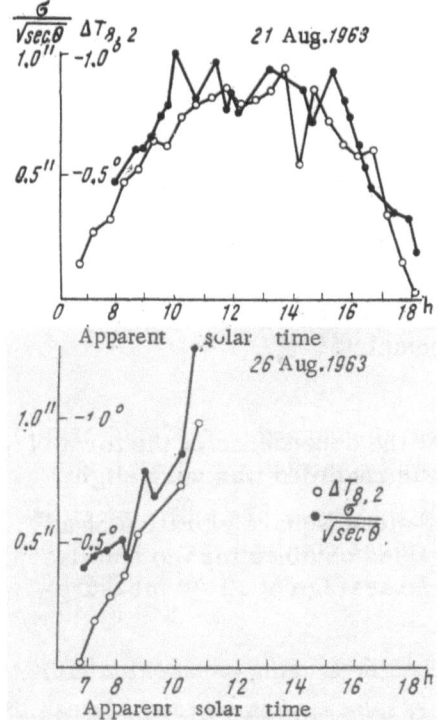

Fig. 3. Comparison of daily variations of root-mean-square vibration of solar-limb image and difference between mean air temperatures at 8 and 2 m in 1963.

For a homogeneous path, when (1) is satisfied, the normed spectral density of arrival-angle fluctuation from a point source has the form

$$\frac{W(f)}{\sigma_\varphi^2} = 0.045 \sin^2 \frac{\pi df}{v_\perp} \left(\frac{d}{v_\perp}\right)^{-5/3} f^{-8/3}, \qquad (8)$$

where v_1 is the wind-velocity component that is normal to the ray.

It is more convenient to represent the frequency spectrum in the dimensionless coordinates $fW(f)/\sigma_\varphi^2$ and $fd/v_\perp$ [15], as we did in processing the obtained data:

$$\frac{fW(f)}{\sigma_\varphi^2} = 0.045 \sin^2 \pi \frac{df}{v_\perp} \left(\frac{df}{v_\perp}\right)^{-5/3}. \qquad (9)$$

The applicability of formulas (6) and (8) to fluctuations of the central position of an extended source with angular dimension γ depends upon the relationship between the objective diameter d and the value γL, where L is the altitude of turbulent heterogeneities in the refractive index. When $\gamma L < d$, pencils of rays from different parts of an extended source will overlap greatly, and the image of all parts of an extended source is shifted as a whole. But when $\gamma L > d$, the rays from the ends of the extended source will pass at a considerable distance over different routes, and different parts of the source can be displaced independently. In this case, when the displacement is averaged over the entire extended source $\Delta\varphi = \int_0^\gamma \delta\varphi(\gamma)d\gamma$ the vibration variance will be underestimated, and high frequencies will be missing in the frequency spectrum. This qualitative reasoning is confirmed by calculations of fluctuations of the arrival angle from an extended source [16].

The height of the layer of turbulent heterogeneities responsible for vibration can be estimated as

$$L \sim \frac{d}{\gamma}, \qquad (10)$$

where γ is the dimension of the extended section in averaging over which a decrease in the vibration variance begins to be observed.

Measurement Results

Figure 1 shows vibration spectra, normed for total variance, measured for various section sizes of the solar-limb image whose displacement was recorded. The sections were 9, 36, 72, 144, and 280". The measurements were made at solar zenith angles of 50 to 30°. From 10 to 19 spectra were obtained for each γ value. Each individual spectrum was reduced to dimensionless form, i.e., represented in the coordinates $fW(f)$ and $fd/v_\perp$. For $v_\perp$ we used the mean-wind component perpendicular to the ray at 70 m as measured on a 70-meter meteorological tower. Then, all of the spectra for one section size were averaged. These

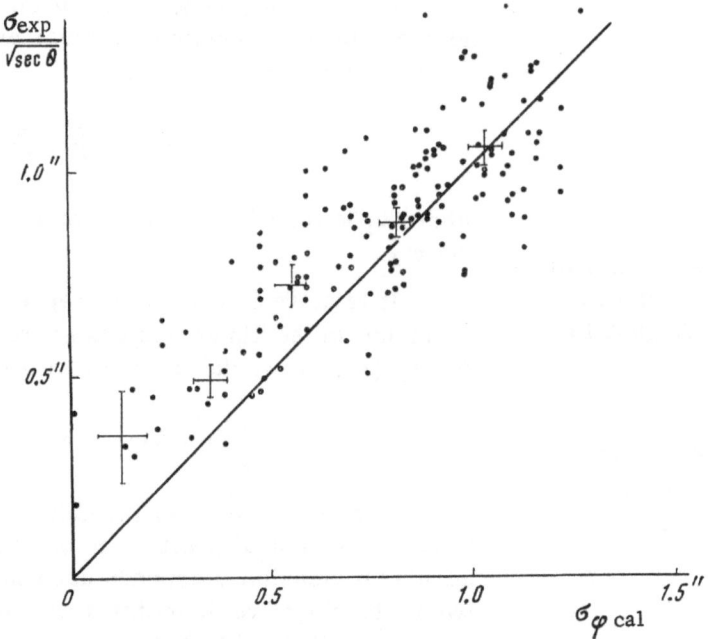

Fig. 4. Comparison of calculated and measured root-mean-square vibration of solar-limb image for 1963 measurements at Tsimlyansk. The crosses denote the group mean and the straight line indicates complete agreement.

averages are also shown in Fig. 1, from which it can be seen that the dependence of the form of the spectrum upon the size of the section whose displacement was recorded was very slight.

Figure 2 shows experimentally obtained ratios of the root-mean-square vibration of an extended-source image σ_{φ_γ} to that of a point source σ_{φ_0}. The values obtained for the smallest measurable section $\gamma = 6''$ were used for σ_{φ_0}. Each point is the average of 12-20 measurements.

The data in Fig. 2 make it possible to estimate the layer height L using expression (10).

When d = 5 cm and $\gamma \sim 1'$, we find that L was several hundreds of meters for the weather conditions under which the vibration measurements were made.

The graphs in Figs. 1 and 2 show rather convincingly that under the weather conditions of our measurements, the point-source theory is easily used to calculate the vibrations of an extended source of up to a few angular minutes in size.

The principal measurements — about 200 variances — were made for $\gamma = 20''$. Figure 3 shows examples of the daily variations of the root-mean-square vibration σ_φ. The difference between the mean air temperatures at 8 and 2 m, as measured on a 12-meter tower near the telescope, is given there for comparison. The vertical temperature gradient in the surface layer can be considered a crude characteristic of the rate of fluctuation of the refractive index for light. It is apparent from the graphs that the daily variation of the vibration is a good replica of the daily variation of the temperature gradient in the surface layer. Note that the daily variation of the variance of the vibration of the solar-limb image is very small for real zenith angles. This is explained by the fact that during the morning and evening hours, when the temperature gradient and refractive-index fluctuations are small, the path traversed by a ray in the atmosphere is considerably increased.

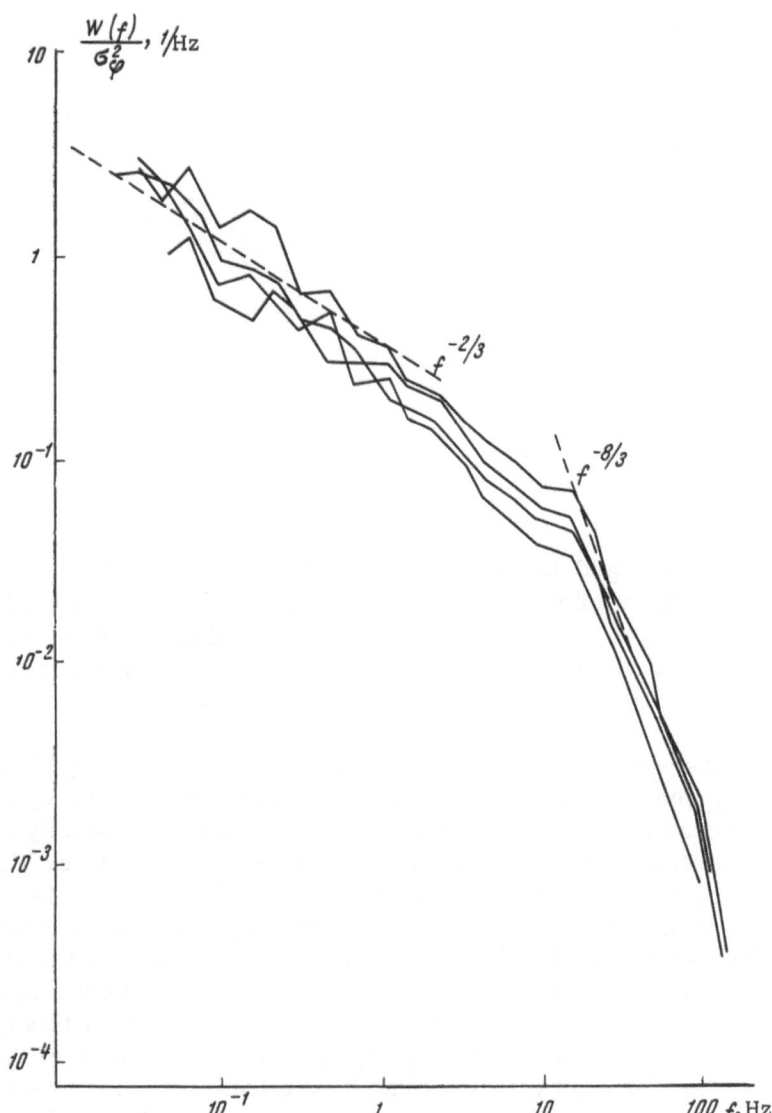

Fig. 5. Examples of frequency spectra of fluctuations in direction of light propagation. The broken lines show the theoretical asymptotes for high and low frequencies.

We used formula (6) to calculate the root-mean-square vibrations for 185 times that corresponded to measurements of this value in Tsimlyansk in 1963. The calculated and measured values are compared in Fig. 4. The calculated correlation coefficients r and the regression coefficients k of $\sigma_{\varphi\,exp}$ on $\sigma_{\varphi\,th}$ using these data are r = 0.83 and k = 1.04.

A series of 30 measurements was made to clear up the question of vibration anisotropy. The horizontal and vertical limb images were projected alternately onto the slit of the instrument using the right-angle prism. It is most interesting to make these measurements during the morning and evening hours — with large zenith angles. In our measurements, the solar zenith angles were from 75 to 60°. The measurements showed that the vertical and horizontal vibration variances were the same within our error limits (10%).

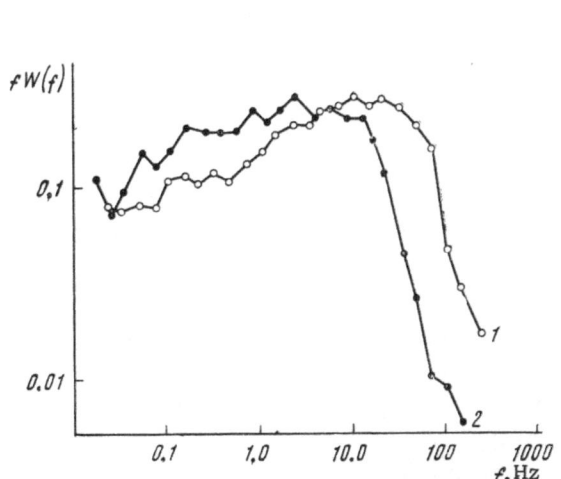 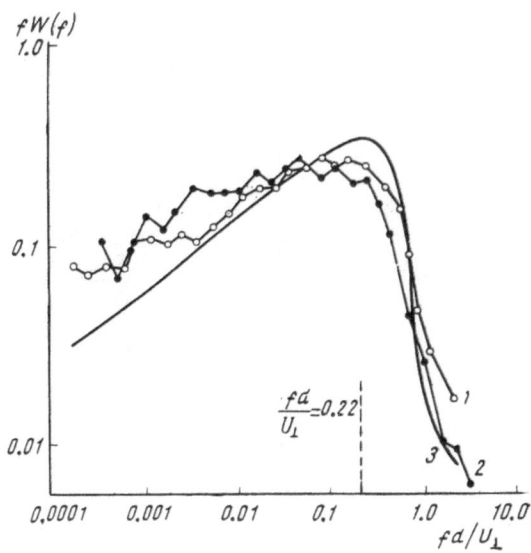

Fig. 6. Comparison of averaged dimensionless frequency spectra of vibrations for
wind velocities (1) $v_\perp$ = 2.5 m/sec and (2)
$v_\perp$ = 7 m/sec measured at 70 m.

Fig. 7. Comparison of shapes of theoretical (3) and measured (1, 2) spectra in dimensionless coordinates. The broken line
indicates the theoretical peak of the dimensionless spectrum.

The measured frequency spectra of the vibrations are compared with the theoretical in
Figs. 5-7. Figure 5 gives examples of individual frequency spectra in the coordinates $W(f)$
and f. There, the broken line shows the asymptotes of the spectral density at low and high
frequencies from theoretical formula (8): $W(f) \sim f^{-8/3}$ for $fd/v_\perp \gg 1$ and $W(f) \sim f^{-2/3}$ for
$fd/v_\perp \ll 1$. Figure 6 gives the mean spectral densities $fW(f)$ for two groups of spectra. The
first group (20 spectra) was obtained for wind velocities $v_\perp$ = 6-8 m/sec; the second (12 spectra), for $v_\perp$ = 2-3 m/sec (the wind velocity was measured at 70 m). All of the spectra were
normed for the total variance. The increase in the contribution of high frequencies to the
energy spectrum when the wind velocity increases is easily seen from this graph. Figure 7
shows the same two averaged spectra, but $fd/v_\perp$ is plotted on the axis of the abscissas, i.e.,
these spectra are shifted for wind velocity. The theoretical frequency spectrum for a homogeneous path and a point source is given there for comparison. For the high frequencies, the
agreement between the calculated and measured spectra is entirely satisfactory; for low frequencies, the experimental spectral density is slightly above the calculated.

Summary

1. Under conditions of well-developed atmospheric convection, which occur in the daytime over dry land in the summer in clear and slightly cloudy weather, fluctuations in the
arrival angle of light from extended sources with dimensions of up to several angular minutes
do not differ appreciably from arrival-angle fluctuations from a point source.

2. Under these weather conditions, vibration is determined by turbulent fluctuations of
the refractive index in a layer a few hundred meters high.

3. The root-mean-square vibrations of solar-limb images under these conditions do not
exceed σ_φ = 1.5" (at the zenith) for objective diameters d $\geq$ 5 cm.

4. The daily variation of the root-mean-square vibrations (reduced to the zenith) provides a fairly good replica of the daily variation of the vertical mean-temperature gradient in
the surface layer near the telescope.

5. Comparison of our measurements with theory confirms that the variance and spectral density of fluctuations of the arrival angle of light that has passed through an atmospheric layer can be predicted with about 30% accuracy using known meteorological parameters (pressure and mean temperature and wind-velocity gradients) for the ray path and V. I. Tatarskii's theory of fluctuation phenomena in wave propagation in a turbulent medium.

LITERATURE CITED

1. V. M. Bovsheverov and M. A. Kallistratova, Astron. Zh., Vol. 49, No. 3 (1964).
2. V. I. Tatarskii, Theory of Fluctuation Phenomena in Wave Propagation in a Turbulent Atmosphere [in Russian], Sections 3, 7, 10, and 13, Izd. Akad. Nauk SSSR (1959).
3. N. V. Bystrova and A. N. Demidova, Izv. GAO, Vol. 22, Issue 4, No. 169 (1961).
4. Telescopes, G. P. Kuiper and B. M. Middlehurst, eds., University of Chicago Press (1960).
5. I. G. Kolchinskii, in: Optical Instability of the Earth's Atmosphere [in Russian], Nauka (1965).
6. N. V. Bystrova and A. N. Demidova, Soln. Dannye, Vol. 12, No. 3 (1961).
7. A. N. Demidova and N. V. Bystrova, Izv. GAO, Vol. 23, Issue 2, No. 173 (1963).
8. N. V. Bystrova and A. N. Demidova, in: Optical Instability of the Earth's Atmosphere [in Russian], Nauka (1965).
9. I. Dommanget, Ciel et Terre, Vol. 80, Nos. 7−8 (1964).
10. L. K. Zinchenko, in: Optical Instability of the Earth's Atmosphere [in Russian], Nauka (1965).
11. Handbook of Geophysics, Macmillan, New York (1960), p. 13.
12. L. R. Tsvang, Izv. Akad. Nauk SSSR, Ser. Geofiz., No. 8, p. 1252 (1960).
13. A. M. Obukhov, Izv. Akad. Nauk SSSR, Ser. Geofiz., No. 9 (1960).
14. L. R. Tsvang, Izv. Akad. Nauk SSSR, Ser. Geofiz., No. 10 (1963).
15. V. M. Bovsheverov, A. S. Gurvich, and M. A. Kallistratova, in: Optical Instability of the Earth's Atmosphere, Nauka (1965).
16. M. A. Kallistratova and A. I. Kon, Izv. Vysshikh Uchebn. Zavedenii, Radiofiz., Vol. IX, No. 6 (1966).

VIBRATION OF STAR IMAGES IN TELESCOPES AS
A FUNCTION OF ZENITH DISTANCE

I. G. Kolchinskii

Determination of the dependence of star-image vibration upon zenith distance (z) is of interest for the following reasons:

1. From theory it follows that star-image vibrations or fluctuations in the arrival angles of light rays must increase in proportion to $L^{0.5}$, where L is the length of the path traversed by the ray in the atmosphere ($L \approx L_0 \sec z$, where L_0 is the path length at z = 0°) [1, 2]. It is important to establish how accurately this relationship works under real conditions.

2. Observations of star-image vibrations at different zenith distances are essential in astroclimatology. In this case, curves showing vibration as a function of z differ, generally speaking, at different locations and on different nights. An explanation of the causes of this difference would be valuable for a study of the effect of locality and weather conditions on astroclimates.

3. The propagation conditions in the atmosphere for various kinds of radiation generated by terrestrial sources, particularly lasers, has gained more and more importance in recent years. It is well known that fluctuations of the optical parameters of the atmosphere destroy such a valuable property of laser beams as coherence. Study of the effect of these fluctuations for different values of z, therefore, is of immediate practical importance [3].

Analysis of Observations at Various Observatories

Below is a summary of the principal results of image-vibration observations at various zenith distances and at different locations. Before analyzing this table, let us make one general comment.

The theory of electromagnetic-wave propagation in media with randomly fluctuating refractive indices leads to a formula according to which the root-mean-square fluctuations of the arrival angles of light rays in the atmosphere increase in proportion to $L^{0.5}$. This law follows from the fact that the characteristic scale of refractive-index heterogeneities is assumed to be very small in comparison with the path length L, i.e., the number of heterogeneities on a beam path is assumed to be very large. Obviously, this circumstance must, as a rule, occur for a particular state of the earth's atmosphere that is realized fairly often under ordinary astronomical observation conditions.

The spectrum of scales of turbulent heterogeneities in the atmosphere is very wide and encompasses the range from one centimer to tens of kilometers and possibly more. Unfor-

18

TABLE 1

Observer	Location	Observation period, years	Instrument	Result
Visual observations				
M. A. Grachev [4]	Kazan, Engel'gardt Observatory	1906-1907	Meridian circle	For high z, vibration proportional to $(\sec z)^{1/2}$; for medium z, it grew more rapidly
T. A. Banakhevich [4]	Kazan, Engel'gardt Observatory	1913	Equatorial telescope $D = 30$ cm	
A. F. Subbotin [6]	Tashkent, Observatory	1927-1928 Winter Summer	Refractors $D = 24$ cm $D = 16.2$ cm	$p = 0.83$ $p = 0.45$
E. K. Kharadze [7]	Abastumani, Observatory, elevation 1500 m	1932-1933	Refractor $D = 7.8$ cm	Vibration increased in proportion to $\log \sec z$
V. G. Fesenkov and V. S. Sokolova [8]	Alma-Ata (Remizovka)	1947	Refractor $D = 30$ mm $F = 500$ mm with very high power	$p = 0.5$
Coutrez and Bossy [9]	Central Africa, Congo, elevation 2100 to 4550 m	1948	Refractor	Some observations gave $p \cong 1$; others gave $p \cong 0.5$. These data refer to "turbulence angles," obtained by Donjon-Kude method
Photographic Observations				
E. Ya. Boguslavskaya [10]	Moscow, Observatory	1903-1935	Refractor $D = 38$ cm $F = 6.5$ m	$p = 0.51$
E. Przybyllok [5]	Königsberg, Observatory	1922-1929	Refractor $D = 33$ cm	Up to $z = 55°$, vibration approximately constant, after which it grows
B. Strömgren [11]	Copenhagen	1945	Telescope closed, diaphragm with two 90-mm apertures	Increase in σ_α with z slightly slower than $(\sec z)^{0.5}$
V. G. Fesenkov [12]	Alma-Ata, Great Alma-Ata Lake	1951-1952	Horizontal telescope $F = 3$ m	Vibration independent of z up to $z = 80°$
I. G. Kolchinskii [13]	Kiev, Goloseevo, Observatory	1955-1956	Astrograph $D = 40$ cm $F = 5.5$ m	$p = 0.53$
Hansson et al. [14]	Lund	1950	–	σ_α proportional to sec z
N. I. Kucherov [15]	Anapa	1956-1957	AZT-7 telescope $D = 20$ cm $F = 10$ m	Mean value of p (or n_σ in Kucherov's symbols) = 0.75
L. N. Tulenkova [16]	Alma-Ata, Astrophysics Institute Observatory, elevation 1550 m	1956-1957	Refractor $D = 20$ cm $F = 300$ cm	$\sigma_z = 0''.30L^{0.36}$ obtained for 23 trails

TABLE 1 Continued

Observer	Location	Observation period, years	Instrument	Result
N. V. Bystrova [17]	Pulkovo	1957-1958	AZT-7	The star image was split by a prism. Most p values were between 0 and 0.5
O. B. Vasil'ev and V. V. Vyazovov [18, 19]	Kuban', Expedition of Main Astronomical Observatory of USSR Academy of Sciences, Botanik, Krasnodar Region	1960	AZT-7	The formula $\sigma_z = \sigma_0 + k \tan z$, which satisfied the observations, was used for processing
K. G. Dzhakusheva, Yu. I. Glushkov, et al. [20]	Kazakhstan, Assy: 2200 m, Blinkovo: 1115 m, Konur-Olen: 1100 m, Kamenskoe Plateau: 1350 m	1960-1961	Refractor D = 20 cm F = 15.7 m	Although dependence of vibration upon z, on the whole, was observed for all the material, it was not always clearly expressed and had different forms
K. I. Kozlova [21]	Southern Kazakhstan, Blinkovo — 1115 m	1960-1961	AZT-7	Clearly expressed dependence upon z not observed. Vibration at zenith sometimes greater than at comparatively high z
O. P. Vasil'yanovskaya [22]	Tadzhikistan, Iskander-Kul' — 2260 m	1960-1963	AZT-7	Author used $\sigma_z = \sigma_0 + k \tan z$, which was in rather good agreement with observations. Representation of mean values by the formula $\sigma_\alpha = \sigma_{\alpha_0} L^p$ gives p values within 0.52-0.60
G. V. Moroz	Kiev, Goloseevo, Observatory	1963	AZT-7	Mean p = 0.51 ± 0.07 for 267 trails (20 nights)

tunately, almost all direct determinations of these scales have been made in the daytime, when the earth's surface was heated by solar radiation, i.e., in the presence of intensive turbulent heat transfer. Under these conditions, as has been shown, for example, in [23], the average dimensions of the turbulent vortices are within 20-200 m. These data were obtained using balanced and attached pilot baloons. It would be interesting to make similar observations at night.

It may be assumed that at night both the intensity and scale of the vortices are decreased, and at sufficiently great zenith distances a light ray will pass through a comparatively larger number of temperature heterogeneities. But in some cases, the frequency of which depends upon the locality and weather conditions (both factors are related), vortices whose scale is not too small in comparison with the path length can exert an influence. Deviations from the $L^{0.5}$ law are possible in this case. For comparison with theory, therefore, it seems advisable to us to average the observation data for different nights and to plot with the obtained points a curve whose parameters can be found by the least-squares method. But when one wants to determine the vibration at the zenith from data for comparatively high z, it is only natural to use results obtained during one night for the extrapolation.

Now let us analyze the results shown in Table 1. It can be seen that conditions under which, on the average, image vibration increases in proportion to $L^{0.5}$ occur rather frequently. This happened, for example, in the visual observations of M. A. Grachev at Kazan', A. F. Subbotin at Tashkent, V. G. Fesenkov, and V. S. Sokolova at Remizovka, and of Coutrez and Bossy in Central Africa, and also in the photographic observations of E. Ya. Boguslavskaya in Moscow, B. Strömgren, in Copenhagen, I. G. Kolchinskii at Goloseevo, N. V. Bystrova at Pulkovo, and finally, of G. V. Moroz at Goloseevo.

It should be noted that if vibrations are represented by

$$\sigma_\alpha = \sigma_{\alpha_0} L^p, \tag{1}$$

where σ_α and σ_{α_0} are the root-mean-square vibrations at zenith distance z and at the zenith, in almost all cases the exponent p will be within the limits $0 < p < 1$.

In our opinion, a shortcoming of the observations of the Kuban' expedition is that, as the authors write, vibration observations by the "trail" method were made "in any wind." In spite of the fact that the AZT-7 telescope was somewhat protected from the wind by the sides of the pavilion and the forest, there is no guarantee that the wind did not affect the vibrations. If the wind had a direct effect on the telescope, the form of the dependence upon z could be distorted considerably, especially if the natural vibration frequency of the instrument were in the range of vibration frequencies under study. The data of the Kuban' Expedition are characterized by an abnormally rapid vibration increase in the near-zenith region, which is not at all in agreement with formula (1), since sec z normally changes very slowly near the zenith.

The results obtained by the author at Goloseevo in 1955-1956, in which p = 0.53 (±0.06), were confirmed by G. V. Moroz in 1963 using an AZT-7 telescope. For Goloseevo, she obtained p = 0.51 (±0.07). Moreover, she discovered a dependence of p upon azimuth. For stars observed in the south, p = 0.41 (±0.06), while for northern stars, p = 0.65 (±0.10). Western and eastern stars gave p = 0.48 (±0.09) and p = 0.23 (±0.05), respectively.

In earlier papers [13, 24] we considered some circumstantial evidence that vibrations increased in proportion to $L^{0.5}$. Thus, the ratio of vibrations at zenith distances z_1 and z_2 must be proportional to $(\sec z_1 / \sec z_2)^{0.5}$. Of course, if this relationship holds for vibrations at given z_1 and z_2, it does not follow that the $L^{0.5}$ law is valid over the entire range of zenith distances. The condition in question is necessary but hardly sufficient for this. The "Catalog of Astroclimatic Characteristics" for 1953, which was published in Novisibirsk [25], gives turbulence angles t_z'' for various stations in Azerbaidzhan at zenith distances of 45 and 70°. We compared the mean values of the ratios t_{70}'' / t_{45}'' for each of these stations:

Station	Mean value of t_{70}'' / t_{45}''	Number of observations
Pirkuli	1.73	151
Altyagach	1.40	39
Dedyugenesh	1.59	76

The values of the ratios $\sec 70° / \sec 45°$ and $(\sec 70° / \sec 45°)^{0.5}$ are 2.07 and 1.44, respectively. It is apparent that at all three points the value of t_{70}'' / t_{45}'' was considerably closer to the former number than to the latter. This indicates that the law of increase of t_z'' with z, at least in the range $45° \le z \le 70°$, is close to the law of $(\sec z)^{0.5}$.

All of the above allows it to be assumed that, since image vibration, on the average, often increases in proportion to $L^{0.5}$, there are atmospheric conditions under which refractive-index heterogeneities are rather small in comparison with the path length of a ray in the layer. But situations are possible in which considerable phase fluctuations can occur on the beam path that mask the effect of small and numerous fluctuations that lead to the $L^{0.5}$ law. Let us list some possible causes.

1. The effect of varying relief together with weather conditions (wind).

a) The effect of relatively major relief, i.e., the presence of unusual conditions on mountain slopes, in valleys, canyons, etc. Wind can form vortex and wave systems on the lee side of mountains. Ascending currents are possible along mountain slopes, and separate systems of strong winds can be formed in valleys and canyons.

The deviation from the $L^{0.5}$ law at Alma-Ata and Abastumani may be explained by these factors.

b) The effect of minor relief: individual hills, buildings, vegetation. With a sufficiently strong wind, vortex systems can be formed on the lee side of obstacles. Such irregularities of relief increase the nonuniformity of relatively large-scale temperature fields, i.e., they disturb the isotropic nature of fluctuations. This may be related to the fact that the erection of large buildings near observatories reduces image quality [26].

c) The effect of pronounced differences in the forms of the relief and the underlying surface in different directions, for example, sea and dry land, forest and city, etc. Since, to find the dependence of vibration upon z, observations are of necessity made at different azimuths, the differences can cause substantial deviations from the $L^{0.5}$ law.

2. The effect of the microclimate of the pavilion. The difference between the inside and outside temperatures of the pavilion can produce temperature-heterogeneous vortex systems near the window.

3. When observing stars near the horizon, large-scale convective jets near the surface, which are to a great extent temperature-heterogeneous, can exert an influence.

4. Wave motions on internal atmospheric interfaces. Formation of a thin vortex-filled boundary layer in which regular wave motions can occur is highly probable in this case. It should be borne in mind that if the perturbing layer is thin, the random phase advance in it will be proportional to the path length of a ray in that layer, i.e., $L_0 \sec z$, where L_0 is the layer thickness. Similarly, the root-mean-square difference between phase fluctuations at two points on the surface of a wave that has passed through the layer will be changed, so that σ_α will be proportional to L rather than $L^{0.5}$.

5. The local nature of turbulence, as observed during airplane and helicopter flights, can also cause deviations from the $L^{0.5}$ law. P. A. Vorontsov and M. A. German, for example, who studied atmospheric turbulence on the Simferopol'-Yalta route with instruments that recorded the g forces undergone by a helicopter, noted that "in general the turbulence was always local and was observed as individual centers" [27].

In the helicopter observations, the scale of turbulent heterogeneities that affected g forces was about 130 m for a bumpy-air zone about 8 km long. Similar observations have been made by N. Z. Pinus [28]. According to Pinus, the size of the turbulent centers varied from 50-100 to 300-500 m vertically and from 1-2 to 4-5 km horizontally. If the distribution of heterogeneities that cause optical instability is also localized, optical perturbations will occur as individual centers of turbulence pass over the observation point. Since vibrations are not observed simultaneously at different z values, deviations from the $L^{0.5}$ law in different parts of the sky is quite probable.

An over-all picture of the vibrations as a function of z can be obtained by adding the effect of the entire atmosphere to the effect of single layers or "centers" that can distort the form of the dependence of σ_α upon L. The relative roles of the factors in disturbing the optical stability of the atmosphere can be determined only by comparing long-term series of observations at different locations.

Remarks on O. B. Vasil'ev's Paper

In his paper "Star Vibration as a Function of Zenith Distance," in "Optical Instability of the Earth's Atmosphere" [18], O. B. Vasil'ev expresses the opinion that the theory of sound-wave and ultrashort radio wave propagation in a medium whose refractive index fluctuates randomly, which was developed by V. A. Krasil'nikov and used by him to explain fluctuations in the arrival angles of light rays in telescopes [1, 29], has no relationship to the phenomenon of star-image vibration. On this basis, O. B. Vasil'ev concluded that our results in [24, 30] are not, as we had assumed, a confirmation of the validity of Krasil'nikov's calculations.

In touching briefly on the questions raised by Vasil'ev in the paper in question, let us say, first of all, that by assuming that V. A. Krasil'nikov's theory is not applicable to image vibration, Vasil'ev is essentially rejecting all of the similar theoretical calculations of S. Chandrasekhar [31], H. Scheffler [32, 33]. V. I. Tatarskii [2], A. M. Obukhov [34], et al. Meanwhile, the conclusions of these authors are confirmed by observations in the optical and radio bands. In particular, they are confirmed by results obtained at Goloseevo in studies of the vibration autocorrelation functions, the correlation between vibrations of the images of stars that are at small angular distances, the dependence of vibration upon objective diameter, and others. Thus, the theoretical data are confirmed not only by studies of image vibration as a function of z but also independently.

V. A. Krasil'nikov's comment in [1] that his reasoning did not hold true for so-called "random refractions" which the observer has time to track while keeping the star on the cross hairs of the astrograph was understood by Vasil'ev as meaning that Krasil'nikov thought that his theory did not hold for star-image vibration!

But every astronomer knows well that what we commonly call "vibration" can refer to relatively high-frequency vibrations that the observer does not have time to track while keeping the star on the cross hairs of the astrograph. Putting it more precisely, the observer does not have time to react to these changes in the position of the star, to make corrections, and to change the position of the instrument. This is why V. A. Krasil'nikov correctly assumed that his theory applied not to low-frequency "random refractions" but to vibration proper.

In the beginning of his paper, in the table on p. 41, in which he classifies the effects of heterogeneities of various scales, O. B. Vasil'ev says that vibrations are caused by heterogeneities whose scales vary from D (D is the objective diameter) to values that are 2–3 orders of magnitude greater than D. Evidently, this must be understood as meaning that the dimensions of the heterogeneities are within rather wide limits — from D to 1000D. If $D \approx 50$ cm, then $1000D \approx 500$ m. It is not difficult to see that this is an overestimate, since with a wind velocity of about 10 m/sec, with heterogeneities of this size the vibration "period" would be about 50 sec, so it would be more natural to consider these image vibrations as "random refractions." With a lower wind velocity, the "period" would be even greater.

On the other hand, if we take 100D as the upper scale limit, the heterogeneity dimensions would be ~50 m at D = 0.5 m. Even in this case, for a vertical layer thickness of the order of 1000 m, at a zenith distance of 60° a ray will pass, on the average, through 2000/50 = 40 heterogeneities. But 50-100 m is evidently the upper size limit of heterogeneities that cause vibration. Heterogeneities with dimensions on the order of the objective diameter make a considerable contribution to image vibration, and there will be thousands of such heterogeneities on a ray path. Thus, in this case, the "law of $N^{1/2}$," where N is the number of heterogeneities on the ray path, must be the determining one.

Also entirely incomprehensible is O. B. Vasil'ev's statement that the law of large numbers can be applied to these heterogeneities only when they have spherical symmetry. Since the theory considers random phase advances with passage through heterogeneities, they can have any form. It is important only that the phase vary slightly over the period of the light wavelength λ [2]. For some reason, Vasil'ev cites Ellison's paper [35], in which it is assumed that vibration can be explained by refraction by large heterogeneities. He writes: "Since refraction theory is well-known, it is unnecessary to write down any relationships." But the theories of V. A. Krasil'nikov, B. Chandrasekhar, et al., are essentially refraction theories of vibration, since they consider wave propagation as an approximation of geometric optics where the wavelength λ is considerably smaller than the characteristic scale of the heterogeneities. In these same theories, when diffraction effects are taken into account, calculation shows that the formulas describing phase-fluctuation phenomena (and, therefore, image vibration) have the same form as in the geometric-optics approximation when diffraction is considered [34].

Which refraction theories is he talking about? The fact that the vibration observations during the Kuban' Expedition can be described by a formula of the form $\sigma_z = \sigma_0 + k \tan z$ certainly does not mean that the term with $\tan z$ is a result of refraction by heterogeneities in the immediate vicinity of the telescope objective. Normal astronomical refraction at $z \leq 75°$ is, in fact, proportional to $\tan z$, but in the case of vibrations we are concerned with fluctuations in ray direction, which are due to differences between phase fluctuations at different points on the objective.

It should be emphasized that the existing theories of image vibration and star scintillation that take into account density fluctuations in a propagated electromagnetic wave are so physically concrete and founded that it would be surprising if observations did not confirm the theory, if only in general outline.

LITERATURE CITED

1. V. A. Krasil'nikov, Dokl. Akad. Nauk SSSR, Vol. 65, No. 3, p. 294 (1949).
2. V. I. Tatarskii, Theory of Fluctuation Phenomena in Wave Propagation in a Turbulent Atmosphere [in Russian], Izd. Akad. Nauk SSSR (1959).
3. A. I. Kon and V. I. Tatarskii, Izv. Vysshikh Uchebn. Zavedenii, Radiofiz., Vol. 8, No. 5, p. 870 (1965).
4. T. A. Banakhevich, Three Studies on Refraction Theory [in Russian], Kazan' (1915).
5. E. Przybyllok, Astron. Nachr., No. 237, p. 5681 (1930).
6. A. F. Subbotin, Tr. Tashkentskoi Obs., Vol. 3, p. 79 (1930).
7. E. K. Kharadze, Byull. Abastum. Obs., Vol. 21, p. 111 (1937).
8. V. G. Fesenkov, Izv. Akad. Nauk Kaz. SSR, Astron. Fiz., No. 1 (1947).
9. R. Coutrez and L. Bossy, Ann. Obs. Roy. Belg., Ser. 4, Vol. 6, No. 5 (1954).
10. E. Ya. Boguslavskaya, Astron. Zh., Vol. 15, No. 5–6, p. 450 (1938).
11. B. Strömgren, Objectiv Registering of den astronomiske Luftuhro, Mat. Tidsskr. Forste Del., Vol. 15 (1945), dedic. to N. E. Norlund.
12. V. G. Fesenkov, Izv. Astrofiz. Inst. Akad. Nauk Kaz. SSR, Vol. 1, No. 1–2, p. 23 (1955).
13. I. G. Kolchinskii, Izv. GAO Akad. Nauk Ukr.SSR., Vol. 3, No. 2, p. 27 (1961).
14. N. Hansson, H. Kristenson, F. Nettelblad, and A. Reiz, Ann. Astrophys., Vol. 13, p. 275 (1950).
15. N. I. Kucherov, Transactions of Conference on Astronomical Scintillation [in Russian], Moscow, 18–20 June 1958, Izd. Akad. Nauk SSSR (1959), p. 183.
16. L. N. Tulenkova, Izv. Akad. Nauk Kaz. SSR, Vol. 7, p. 74 (1958).
17. N. V. Bystrova, Transactions of Conference on Astronomical Scintillation [in Russian], Moscow, 18–20 June 1958, Izd. Akad. Nauk SSSR (1959), p. 155.

18. O. B. Vasil'ev, in: Optical Instability of the Earth's Atmosphere [in Russian], Nauka (1965), p. 40.

19. O. B. Vasil'ev and V. V. Vyazovov, Izv. GAO SSSR, Vol. 22, Issue 5, No. 170, p. 144.

20. K. G. Dzhakusheva, Yu. I. Glushkov, et al., Tr. Astrofiz. Inst. Kaz. SSR, Vol. 4, No. 5 (1963).

21. K. I. Kozlova, Tr. Astrofiz. Inst. Kaz. SSR, Vol. 4, p. 49 (1963).

22. O. P. Vasil'yanovskaya, Byull. Inst. Astrofiz. Akad. Nauk Tadzhik SSR, Nos. 39-40, p. 47 (1965).

23. T. V. Bonchkovskaya and V. G. Nikitin, in: Study of Atmospheric Heat Transfer [in Russian], Nauka (1964), p. 86.

24. I. G. Kolchinskii, Astron. Zh., Vol. 23, No. 3, p. 350 (1952).

25. Catalog of Astroclimatic Characteristics for 1953 [in Russian], No. 1, Novosibirsk (1964).

26. D. Ya. Martynov, A Course in Practical Astrophysics [in Russian], Fizmatgiz (1960).

27. P. A. Vorontsov and M. A. German, Trudy GGO im. A. I. Boeikov, No. 171 (1965).

28. N. Z. Pinus, Izv. Akad. Nauk SSSR, ser. fiz. atmosf. i okeana, Vol. 1, No. 3, p. 266 (1965).

29. V. A. Krasil'nikov, Izv. Akad. Nauk SSSR, ser. geogr. i geof., Vol. 13, No. 1. p. 33 (1949).

30. I. G. Kolchinskii, Astron. Zh., Vol. 34, No. 4, p. 38 (1957).

31. S. Chandrasekhar, Monthly Notices Roy. Astron. Soc., Vol. 112, No. 5, p. 475 (1952).

32. H. Scheffler, Astron. Nachr., Vol. 282, No. 5, p. 193 (1955).

33. H. Scheffler, Astron. Nachr., Vol. 284, No. 5, p. 227 (1958).

34. A. M. Obukhov, Izv. Akad. Nauk SSSR, seriya geofiz., No. 2, p. 155 (1953).

35. M. A. Ellison, in: Proc. Symposium on Astronom. Optics and Related Subjects, Manchester, 1955; Z. Kopak, ed., Amsterdam (1956), pp. 293-299 in the Amsterdam edition.

DISTRIBUTION OF WAVE-FRONT DISTORTIONS
THAT CAUSE STAR-IMAGE VIBRATION

Sh. P. Darchiya and N. P. Esikov

As is well known [1], the propagation velocity of light is reduced by a factor of n in a medium with refractive index n. If light in the atmosphere passes through some heterogeneity with thickness H, the wave front will be deformed and the deformation h will be expressed by the formula h = H(n − 1).

Not a plane wave front but a considerably distorted one with a very complex undulating surface will enter the telescope objective. It has been established that the less linear the deformation scale the smaller the deformation index h.

Wave-front deformations of different scales play different roles in the construction of a star image in the focal plane of the objective. For example, deformations that are much smaller than the telescope diameter distort the diffraction pattern of stars — they break it up and blur it [2, 3]. Deformations of length D cos θ, where D is the telescope diameter and θ is the angular amplitude of the vibration, cause rapid shifting of the entire diffraction pattern, as will be shown below. The form of the deformed wave front entering the objective (or at least the form of a cross section of this front) and how deformations affect the structure of the image in the telescope are very important problems for both theory and practice.

Geometric Interpretation of Image Vibration

We find in front of the telescope objective by the least-squares method a median plane Q that passes through a deformed front that causes blurring of the diffraction pattern. For the sake of brevity, the section of the front approximated by the plane will be called plane Q. When plane Q is parallel to the plane of the objective P for a certain period of time the star image will be observed at point F on the optical axis and image vibration will be absent. If at time t plane Q is at an angle θ to plane P, the star image will be shifted relative to point F. If, for example, plane Q occupies at time t_1 position Q_1 (Fig. 1), the image will be observed at point M_1 on the photographic plate P', which is in the focal plane of the telescope. If plane Q at succeeding time t_2 occupies position Q_2, the image will be shifted to point M_2, which is on the other side of point F. The image displacement from the optical axis expressed in arc measure will be equal to the angle of plane Q relative to the plane of the objective.

Thus, variation in the inclination of the plane from one moment to another causes image vibration, and this inclination is equal to the vibration amplitude in arc seconds. When a star is photographed at the focus of a fixed telescope, a trail is obtained on the film.

In this paper, the vibration characteristic will be $\sigma_i = A_{av} - A_i$, where $A_{av} = \Sigma A_i / N$ is the average reading in measuring a trail at N points and A_i is the reading for the image position at time t_i.

26

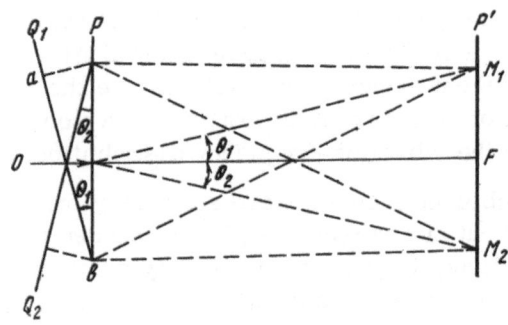

Fig. 1. Structure of star image when section of plane Q is at angle to objective plane.

Let us consider in detail the formation of the trail curve in Fig. 2. Let plane Q_1 be parallel to plane P at time t_1, and let the image run over the plate P' from left to right. Then, at time t_1 the image will be observed at point k_1 on the star trail. If during the succeeding moments plane Q_1 changes its inclination as shown by the broken lines in Fig. 2, the image on the plate will describe curve segment K_1M_1. When the inclination reaches its maximum Q_2, the trail will be found at point M_1. Then if plane Q begins to reduce its angle relative to plane P and this angle is zero at time t_3, the image will describe curve segment M_1K_2 during the time t_3-t_2.

Thus, during the time t_3-t_1 wave-front segments with dimension $D\cos\theta$ have a common orientation relative to the objective plane, but for the time interval corresponding to segment $K_2M_2K_3$, they will have the opposite orientation (position of plane Q_4). The duration of the orientation changes determines the deformation of the wave front to a greater or lesser extent.

The time distribution of the deformation lengths of a wave front entering the objective can be obtained by counting the groups with the same number of peaks of the same sign.

Distribution of Wave-Front Deformations

In passing through the entire atmosphere, a wave front can encounter several heterogeneities with different characteristics. Each heterogeneity individually will deform the wave front in a particular way, and a segment of the front with the resultant (total) deformation from all heterogeneities will enter the objective. We shall assume that the segment of the front has passed through only one effective heterogeneity. Each segment of the star trail will correspond to a specific segment of the deformed front. Therefore, the distribution of wave-front deformations as calculated from star trails will also be the length distribution of these total heterogeneities as read in the direction of star motion.

Let parallel alignment of plane Q and the objective plane, i.e., amplitude $\sigma_i = 0$ on the star trail or a change in the sign of σ_i, be called event k. The number of these events is represented by a sequence of points $k_1, k_2,...,k_n$ corresponding to the specific times of occurrence of the event.

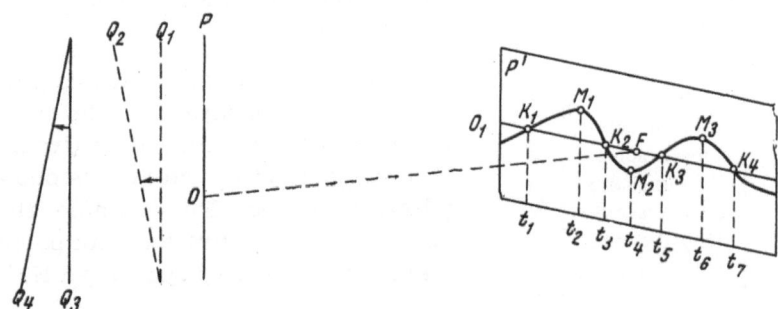

Fig. 2. Formation of star vibration "trail" as inclination of plane Q changes.

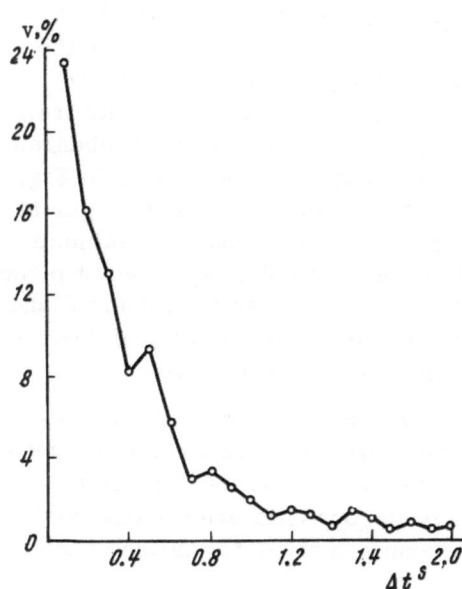

Fig. 3. Length distribution of intervals Δt^s (Novosibirsk, 1962).

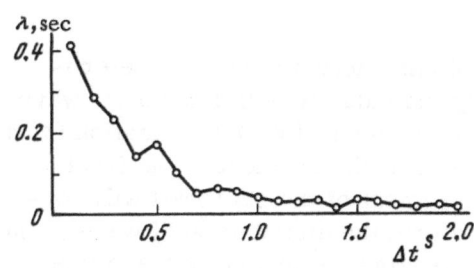

Fig. 4. Distribution of interval densities for 1 sec (Novosibirsk, 1962).

TABLE 1.

Date photographed in July 1962	Star	z^o
1/2	γ Draconis	2
2/3	γ Draconis	2
2/3	II Herculis	2
3/4	γ Draconis	2
4/5	γ Draconis	2
5/6	γ Draconis	2
6/7	γ Draconis	2
8/9	γ Draconis	2
13/14	O Cygni	7

An important characteristic of event k is the length distribution of the random time interval Δt between two adjacent events. This will also be the length distribution of the front deformations, which is equivalent to the vibration–frequency distribution.

If the number of events k were stationary, ordinary, and without aftereffect, i.e., corresponded to steady Poisson flow, the probability density function for the length of the random time interval Δt between adjacent events would be

$$f(t) = \lambda e^{-\Delta t}, \qquad (1)$$

where λ is the flow density (the average number of events per unit time) [4].

In our case, the distribution is obviously more complex, and at hardly any observation can the length–density curve for the random time interval Δt be approximated by (1).

Using the example of nine trails, we shall examine the distribution of Δt from measurements at Novosibirsk in 1962 with an AZT-7 telescope (D = 200 mm, F = 10 m); z^o is the zenith distance. See Table 1.

Since the amplitudes σ_i were measured at $0^s\!.1$ intervals, the length of the time interval $\Delta t = 0^s\!.1 \times n$, where n is the number of peaks between adjacent events.

For this observation point, there are, on the average for nine trails, 71 intervals in a 40–sec time segment, which gives a mean density $\lambda = 1.8$ for 1 sec. The weighted mean interval $m_i = 0^s\!.56$ and the variance $\delta_t = 0^s\!.47$ [4].

Figure 3 shows a graph of the frequency of appearance of intervals of different lengths. The intervals are plotted on the axis of the abscissas and their frequency of appearance v is plotted on the axis of the ordinates.

It is easy to see that this curve illustrates the frequency of appearance of image vibrations with different vibration frequencies (from 1/4 to 5 Hz). In fact, the size of an interval is one-half the image-vibration period. For example, the interval $\Delta t = 0^s\!.1$ corresponds to vibration with period $T_1 = 0^s\!.2$ and, therefore, with frequency v = 5 Hz.

The density distribution (Fig. 4) of intervals of different lengths of a 1-sec segment is of particular interest. It shows the fraction of the total density $(\lambda = 1.8)$ accounted for by an interval of a given length. The densities are defined as

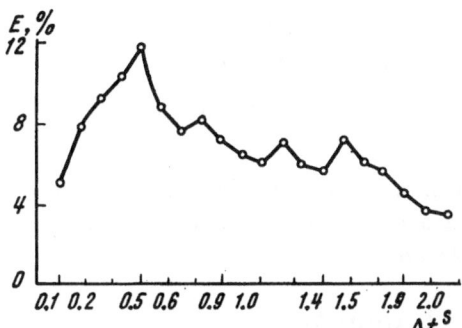

Fig. 5. Distribution of effectiveness of intervals Δt^s by length.

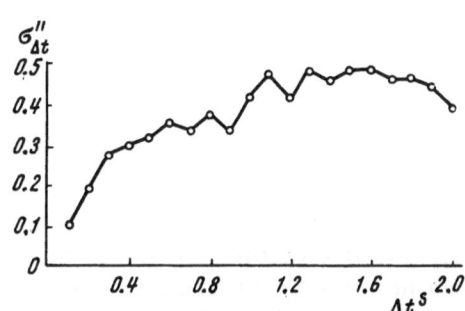

Fig. 6. Vibration amplitude versus Δt^s.

the ratio of the number of appearances of an interval of given length to the total time segment (40 sec).

Determination of the Lower Limit of Wave-Front Deformation Dimensions

Since a star image is not a point source, a wave-front segment whose length is not less than $D \cos \theta$ corresponds to each trail segment that is equal to the diameter of the spurious disk α. If the size of the interval for time Δt^s on the trail is αp, the size of the wave-front deformation for this same time must be not less than $\sum_{i=1}^{p} D \cos \theta$. If we let L be the deformation size and bear in mind that $\cos \theta$ is close to unity, we can write approximately

$$L \geqslant pD. \tag{2}$$

The lower limit of deformation size in space is $L = pD$. For the AZT-7 telescope, $\alpha = 0.034$ mm. The image covers this trail distance (for the stars in question) in $0\overset{s}{.}07$. If we let this value be the exposure time, the value of p in (2) for each time interval Δt^s examined can be approximated by

$$p = \frac{\Delta t^s}{0^s\,07}. \tag{3}$$

Values of p calculated by formula (3) and the lower limits of deformations L that cause vibration in the AZT-7 telescope are:

Δt^s	0.1	0.5	1.0	1.5	2.0	2.5	3.0	3.5	4.0	4.5	5.0
p	1.4	7.1	14.3	21.4	28.6	35.7	42.9	50.0	57.1	64.3	71.4
L, cm	28	142	286	428	572	714	858	1000	1142	1286	1428

Intervals Δt^s with dimensions $0\overset{s}{.}1$, $0\overset{s}{.}2$,...,$5\overset{s}{.}3$ correspond to vibrations with frequencies from 1/11 to 5 Hz. These frequencies result from front deformations whose lower size limits are 28, 142,..., 1428 cm, respectively.

Effective Frequencies

The appearance frequency of a deformation of a particular size by itself says little about the contribution of that deformation to the mean vibration amplitude at a given observation point.

The distribution (Fig. 5) is such that the frequency of appearance on the wave front decreases with an increase in Δt^s. But it is still possible that the contribution of large deformations to the mean is not reduced. It would also be interesting to find out which deformations are most effective for a given point. The effectiveness of a deformation corresponding to part of a time interval Δt^s can be found as the ratio of the sum of all amplitudes obtained in the interval of the deformation to the sum of the amplitudes in the interval Δt^s:

$$E = \frac{\sum_{i} \sigma_{i(k)}}{\sum_{i} \sum_{k} \sigma_{i(k)}}.$$

The total effectiveness for two adjacent intervals Δt^s was used to show more clearly the distribution of effectiveness by interval length Δt^s in plotting the graph (Fig. 5).

Analysis of Fig. 5 shows that the effectiveness increases rather rapidly, has a maximum at $\Delta t^s = 0.5-0.6$, and then decreases slowly.

Thus, for Novosibirsk the interval lengths Δt^s in which maximum effectiveness occurs are $0.5-0.6$, which correspond to a vibration frequency of the order of 1 Hz. The lower size limits of deformations resulting in this frequency are 142-172 cm.

Variation of Vibration Amplitude with Interval Length

In determining the effective lengths of the intervals, we took into account the frequency of appearance of the interval as well as the vibration amplitude. It is also important to show how the mean vibration amplitude $\overline{\sigma}^{''}_{\Delta t}$ varies:

$$\overline{\sigma}^{''}_{\Delta t} = \frac{\sum\limits_{i=1}^{n} \sigma^{''}_{i(\Delta t)}}{n} \, ,$$

where $\sigma^{''}_{i(\Delta t)}$ is a single vibration amplitude obtained in interval Δt^s and n is the number of such amplitudes.

Mean amplitudes are not given for intervals greater than 2.0, since their frequency of appearance is small. The tendency for $\overline{\sigma}^{''}_{\Delta t}$ to increase with an increase in Δt^s from 0.1 to 1.6 is easily seen from the graph in Fig. 6. The decrease in amplitude at $1.6-2.0$ is evidently explained by the fact that for $D = 20$ cm, the angle θ decreases with high L (see Fig. 1).

Thus, image vibration can be explained by inclination of the median plane. An isolated image shift is caused by inclination of a wave-front segment equal to $D \cos \theta$, where $\tan \theta = \sigma/F$.

The frequency distribution of Δt^s reflects the star-image vibration-frequency distribution.

The effective vibration frequencies for the observation periods given above are on the order of 1 Hz.

As the interval length Δt^s increases, i.e., as the vibration frequency decreases, the mean amplitude increases, which reflects an increase in the angle of plane Q to the principal objective plane and, therefore, indicates something about the thickness and density of the total heterogeneities.

The vibration-frequency distribution and the most effective frequencies for observation points are very important in a study of the vibration process and are necessary for designing homing instruments and systems.

We thank Dr. V. S. Stepanov for his valuable comments.

LITERATURE CITED

1. D. Ya. Martynov, A Course in Practical Astrophysics [in Russian], Fizmatgiz (1960).
2. V. N. Tatarskii, Statement on E. A. Blyakhman's Paper, in: Transactions of Conference on Astronomical Scintillation [in Russian], Izd. Akad. Nauk SSSR (1959).
3. S. I. Sorin, Address, in: Transactions of Conference on Astronomical Scintillation [in Russian], Izd. Akad. Nauk SSSR (1959).
4. E. S. Venttsel', Probability Theory [in Russian], Fizmatgiz (1962).

THE PROBABILITY OF OBTAINING GOOD STAR IMAGES WITH SHORT-EXPOSURE PHOTOGRAPHY

V. F. Anisimov, R. N. Berdina, N. N. Nechaeva,
P. V. Nikolaev, and I. P. Rozhnova

Recently, along with the study of astroclimate and the characteristics of star scintillation, projects have been developed and devices created that somehow compensate for the effect of atmospheric instability on the quality of star images. Two trends in the solution of this problem are most clear. The first is the development of schemes and devices for automatic compensation [1]. In a recent paper, N. F. Kuprevich gives a thorough analysis of the possibilities of this course [2]. Very interesting studies of efficient means of compensating for the effect of atmospheric instability on image quality are being conducted at the Main Astronomical Observatory of Pulkovo under the direction of N. F. Kuprevich. The other method of reducing the effect of atmospheric interference in optical observations is to develop so-called indices of image quality [2]. In this case, observations are made at times when the image quality satisfies the observer. Underlying this method is the assumption, which was made long ago by Donjon and Kude, that at some instant it is likely that the atmosphere will quiet down on the line of sight from the telescope objective to the star [3]. The work of Platt, who in 1956 greated a time-selection system for high-quality images [4], and the image-quality indicator of Bray, Loughhead, and Norton [5] are noteworthy.

The results of experimental studies made at the base of the Crimean Astrophysical Observatory to estimate the probability of high-quality star images with very short exposures are given below.

The experimental apparatus consisted of an AZT-7 telescope, a photoelectric indicator for relatively good-quality star images, based on a photoelectric device with a dividing prism for studying star vibration and scintillation [6], and highly sensitive television equipment with a high-speed motion picture camera.

The operation of the apparatus is explained in Fig. 1. The luminous flux from the star, which is gathered by the telescope (AZT-7, D = 200 mm), is divided into two beams by a cube 2. One beam, which contains the greater part of the light, goes through the objective 4 and through the television system to the motion picture camera 19, which has a high-speed shutter (equivalent focus 165 m). The second beam is focused by a stretching lens 3 onto the edge of the dividing prism 8 (equivalent focus 10 m). A semitransmitting mirror 5, an eyepiece 6, and a cross-hair measuring scale 7 are used to estimate the size of the star image and to guide it.

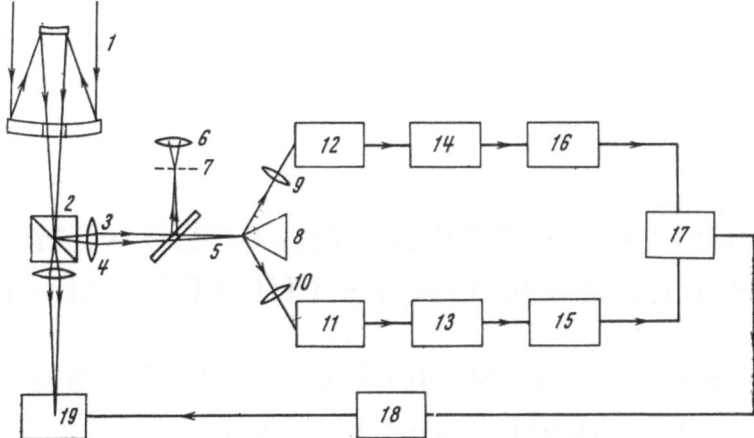

Fig. 1. Block diagram of experimental apparatus.

The dividing prism, in turn, divides the image into two parts. The edge of the prism follows the diurnal motion of the celestial sphere, and the process is scanned about the hour circle. The telescope drive error is less with this type of optical system. The luminous fluxes reflected from the faces of the prism are aimed by Fabry lenses 9 and 10 at the cathodes of two photomultipliers 11 and 12. The photomultiplier signals go through cathode followers 13 and 14 and amplifiers 15 and 16 to a balanced circuit 17. When the star image is quiescent and the luminosity is distributed uniformly over its turbulent disk, the signals in the balanced circuit are balanced. When the image is displaced from the edge of the dividing prism, due to vibration or luminosity redistribution, unbalance occurs, which indicates a reduction in image quality. An output-signal limiter 18 after the balanced circuit is used to set the discrimination level for the high-speed shutter of the motion picture camera 19.

The recordings have shown that the probability of balance and its duration are functions of the telescope aperture, weather conditions, the elevation of the star above the horizon, etc., and that they vary within very wide limits. For a telescope with a 200-mm objective diameter in the autumn, when images are very unstable and their quality, as a rule, is poor, balance durations have varied from 0.1 to 0.005 sec. The number of balances in about 20 recordings varied from one to 10 in the course of one second.

Let us assume that the balance states detected with such an analyzer prism correspond to an improvement in image quality. The validity of this assumption will be determined below. Note that the unbalance state, if it is not due to clockwork errors, indicates conclusively the moments of image-quality impairment. Figure 2 shows a graph of the duration distribution of the balance state. Since there is always a slight underbalance due to photomultiplier noise, this underbalance value plus a small margin was used as the discrimination level, in order to eliminate the effect of sudden noise surges. The graph was plotted using oscillograms with a duration of 10 sec each, on the average. Curve group 1 was plotted from observations at the Crimean Astrophysical Observatory during the autumn-winter period (3 November 1963, Capella). At the moment of observation, the images were poor (on the order of 5"). Curve group 2 is for observations in the spring of 1965 with relatively good image quality (on the order of 1-2"). The data in Fig. 2 allow short exposure times to be selected with greater certainty. It follows, moreover, that reducing the exposure time below some critical value (for a given aperture) is not likely to increase the probability of obtaining good-quality images. An exposure of 1/200 sec (5 msec) can be considered the critical value for the data in Fig. 2, for a 200-mm aperture.

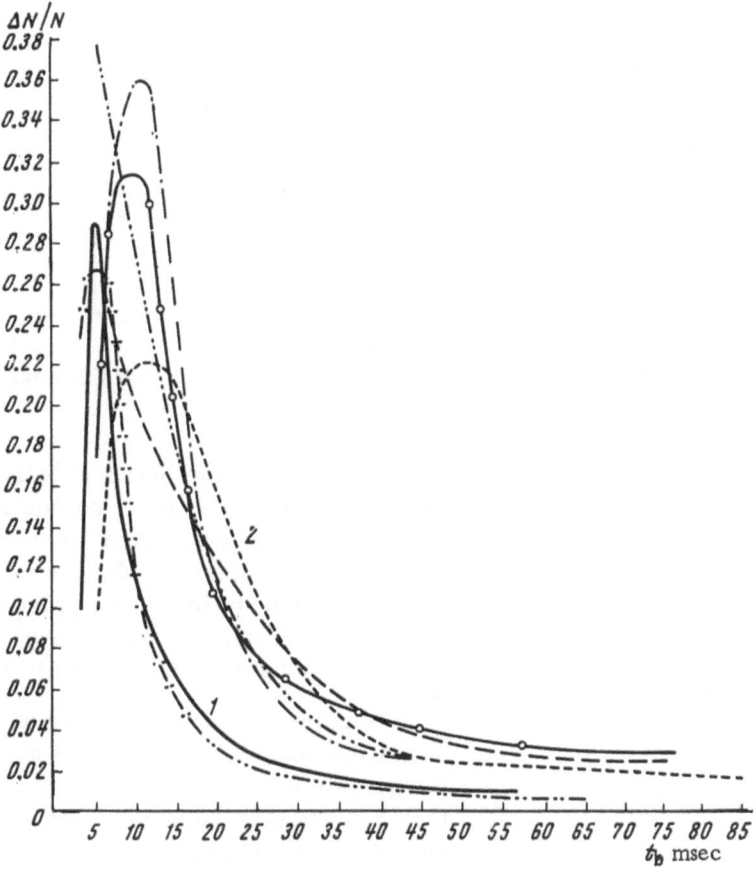

Fig. 2. Distribution of balance state durations.

The television equipment is described in this collection by A. N. Abramenko and V. V. Prokof'eva and elsewhere [7, 8]. It had an electronic shutter, which was specially developed by A. N. Abramenko, that allowed star images to be stored on the target of the image orthicon only for specific time periods (8, 12, 18, 40, 60, and 100 msec). When a given exposure time had elapsed, the brightness amplifier was cut off and the stored image was read. The reading time, which included opening of the camera shutter, photographing the frame, and advancing the next frame, was 1.5-1.7 sec. Therefore, a frame whose exposure was set by the electronic shutter could be obtained every 1.5-1.7 sec.

The electronic shutter could be actuated either manually by a button every 1.5-1.7 sec or by control pulses generated by the output-signal limiter (18 in Fig. 1) when a balance state occurred. The control pulse, discrimination threshold, noise level, etc., are shown schematically in Fig. 3. A control pulse u_c is generated only when the desired signal u_s has become balanced and is equal to the discrimination level u_d.

The images were photographed from the image-orthicon screen in series of about 100-150 frames each alternately by manual control (pushing the button every 1.7 sec) and by pulse control of the electronic shutter. With manual control, of course, there is no way to check image quality. Automatic pulse control assumes the presence (with a certain probability) of quality control. It goes without saying that the series were photographed with identical exposure times and under identical equipment and observation conditions. The alternation of short series, which took an average of 3-10 min on film, enabled us to avoid errors due to variations in transmission, zenith distance, etc.

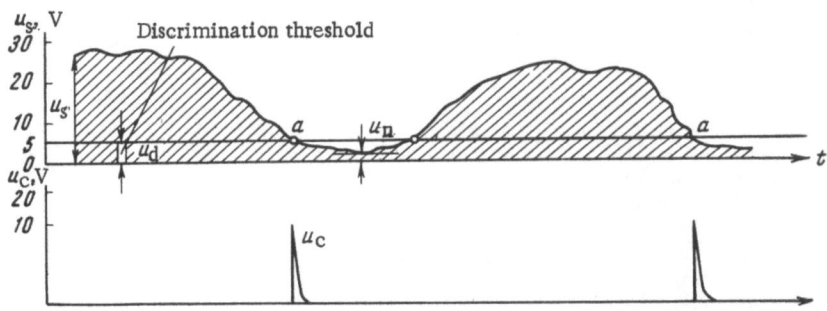

Fig. 3. Illustration of control-pulse generation by photoelectric
pickup.

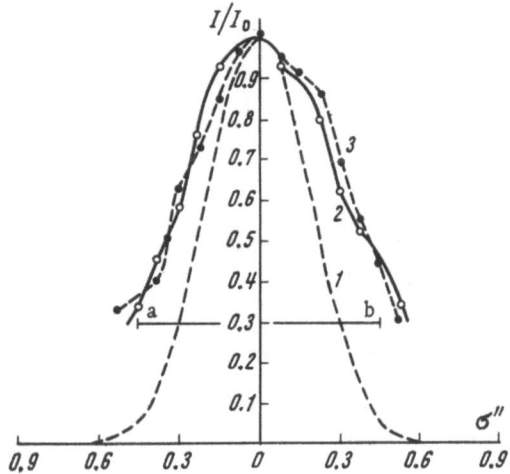

Fig. 4. Photometric profiles of stars.

A chronograph that was triggered by the electronic-shutter control pulses was used several times. It gave the average time intervals for triggering of the electronic shutter manually (approximately 1.5 sec) and by control pulses (1.7–4 sec). The double star β Cygni, whose separation is 34.''6, was used to calibrate the image scale in arc seconds.

Over 16,000 frames were taken in the observation season from May through June 1965. For the given aperture and observation season, satisfactory results were obtained with exposures of not over 8 msec. For 100- and 60-msec exposures, the images showed appreciable averaging (average dimensions of 5–4''). From the 16,000 frames, 10,000 were selected for a statistical estimate of the ratio of good images to the total number for manual and for automatic control. The other 6000 were rejected, either because of the observation conditions (clouds, wind) or overexposure.

The image shape (ideally a circle) and size were taken as the main criteria for image quality. The theoretical radius of the spurious disk for our telescope was about 0.''7. The maximum resolution of the television equipment was 0.''9, so an image was considered good if it were sharp, circular, and within 0.9–1'' in size. For a more objective estimate of image size, about 200 photometric profiles of good images were made on an MF-2 microphotometer. Figure 4 shows photometric profiles of good star images obtained with automatic (curve 2) and manual (curve 3) operation of the electronic shutter. For comparison, curve 1 gives the theoretical energy distribution in the center spot of the diffraction image for a 200-mm aperture. The resolution of the entire recording channel is shown in the scale of the figure by segment ab. Thus, our quality criteria make it possible to select almost-theoretical star images. The results of visual evaluations (on the scale of the mat projection screen) were in good agreement with estimates of the photometric profiles.

The question of brightness as an index of image quality should be approached very carefully. First of all, scintillation exerts a considerable influence with such short exposures. Moreover, the good (in shape and size) images had, as a rule, the average brightness of the entire series, i.e., the series contained bright, irregular (elongated ovals, doubles, etc.),

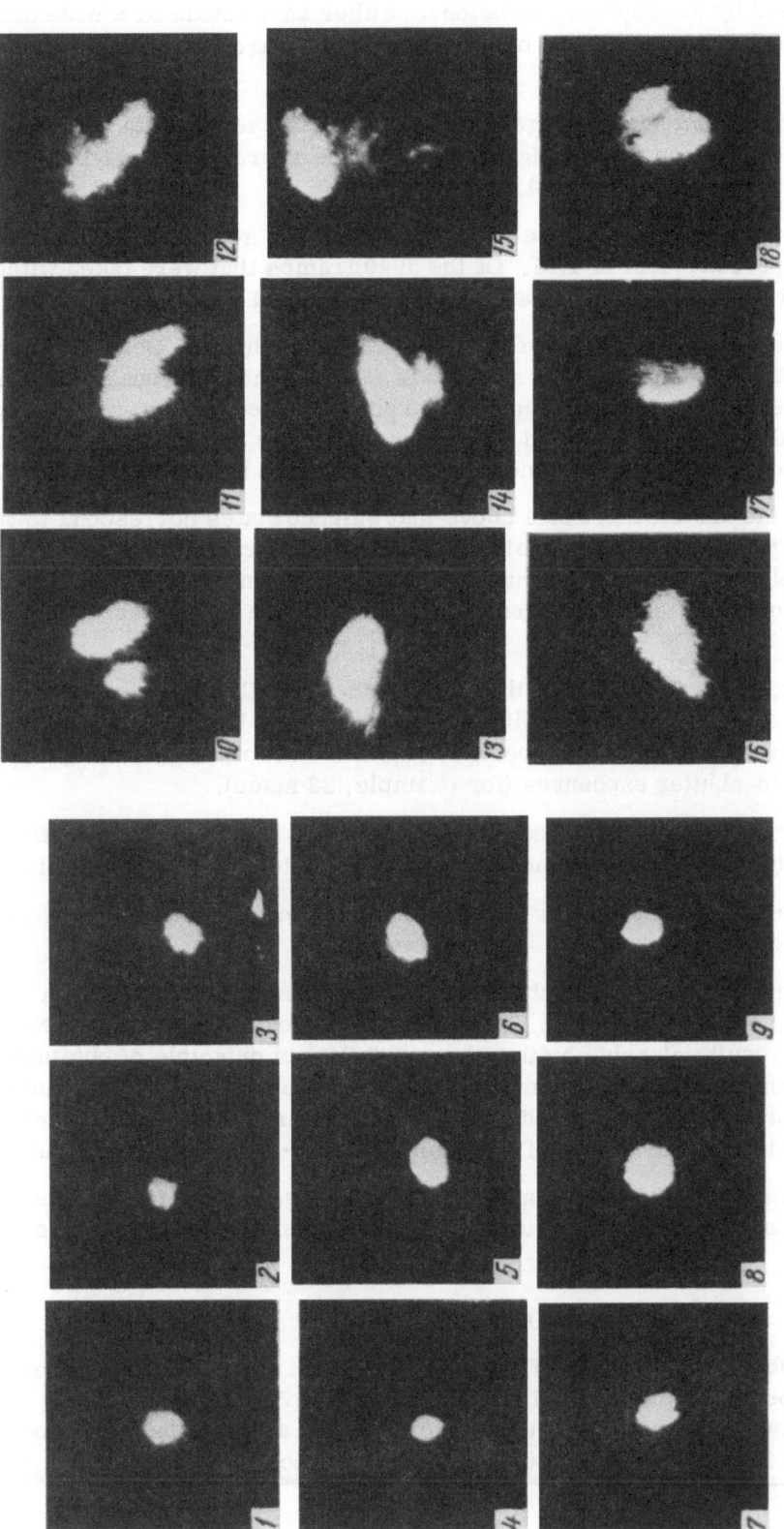

a

b

Fig. 5. Motion picture frames of good (a) and poor (b) star images.

large-size images as well as poor (broken, large) weak images. This is possibly explained by the fact that when a plane light wave that is undistorted by atmospheric turbulence enters the telescope (which then produces a good image), the rays strike the mirror in parallel. When a plane light wave is distorted by atmospheric turbulence, either an antinode or a node can enter the telescope. This reduces or increases the brightness of the image, simultaneously reducing its quality.

Figure 5a shows some typical good images selected from the series with pulse control of the electronic shutter. For comparison, Fig. 5b shows some representative types of poor images obtained with manual and pulse control.

Of the 4830 frames taken with image-quality control, 56 were good images (according to the above criteria), for a percent yield of 1.15. Of the 5040 frames that were taken without quality control, eight good frames were obtained, for a percent yield of 0.16.

Thus, for the given observation season at the Crimean Astrophysical Observatory with a 200-mm aperture and an 8-msec exposure, an average of at least one good-quality image is obtained in every 1000 frames. These 1000 frames can be photographed in 30 min. Quality control increases the probability of obtaining good images by one order of magnitude. One good image can be obtained from as few as 100 frames, and this requires not more than 10 min.

Now let us evaluate the assumption made above that balance states correspond to times when images are improved. Statistical processing of the series of frames made with quality control gives approximately 1% good-quality images. This figure can serve as the likelihood estimate of the above assumption. Therefore, only 1% of the balance states correspond to truly good-quality star images.

As follows from Fig. 2, however, 25% of the total number of balance states had durations of the order of 8 msec, the exposure time. Most likely, there were real improvements in image quality at precisely these moments. This is supported by the fact that not one good image was obtained with longer electronic-shutter exposures (for example, 28 msec).

If this is a valid assumption, then with our method of quality control the theoretical limit for obtaining a good-quality image at the moment of a balance state is 25% (but we obtained 1%).

Perhaps the real yield of quality images with photoelectric control can be increased somewhat by improving the method of control-pulse generation.

It should be emphasized that all of the above data are for a specific observation season, a 200-mm aperture, and an 8-msec exposure. The material cannot be generalized. Nevertheless, the experimental results clearly confirm the possibility in principle of obtaining high-quality images with very short exposures when a photoelectric image-quality indicator is used. It is probable that studies by our method as well as by other methods will enrich our knowledge of the least-studied manifestation of star scintillation — so-called image pulsations.

We thank the management of the Crimean Astrophysical Observatory for permission to work there and A. A. Abramenko and V. V. Prokof'eva of the Observatory for assistance in the observations.

LITERATURE CITED

1.   V. P. Linnik, "Possibility in principle of reducing the effect of the atmosphere on star images," Optika i spektroskopiya, Vol. 3, No. 4, p. 401 (1957).
2.   N. F. Kuprevich, "Some possibilities of using photoelectric and television methods to combat interference from atmospheric turbulence in astronomical observations," Izv. GAO, Vol. 23, Issue 5, No. 175 (1964).

3.   A. Donjon and A. Kude, "Atmospheric agitation," Astron. zh., Vol. 17, No. 1 (1940).

4.   J. R. Platt, "Increase of telescope resolution with time selection and an image forming stellar interferometer," Astrophys. J., Vol. 125, p. 601 (1957).

5.   R. J. Bray and R. E. Loughhead, "Facular granule lifetimes determined with a seeing-monitored photoheliograph," Austral. J. Phys., Vol. 14, p. 10 (1961).

6.   I. P. Rozhnova and Yu. A. Sabinin, Photoelectric Method for Studying Star Image Instability, New Techniques in Astronomy [in Russian], Izd. Akad. Nauk SSSR (1962).

7.   A. N. Abramenko, "An electronic shutter with speeds from 0.005 to 0.1 sec," Izv. KrAO, Vol. 34 (1966).

8.   A. N. Abramenko and V. V. Prokof'eva, "Highly sensitive electronic apparatus for short-exposure photography," Izv. KrAO, Vol. 35 (1966).

ELECTRONIC SYSTEMS FOR SHORT-EXPOSURE
PHOTOGRAPHY OF CELESTIAL OBJECTS

A. N. Abramenko and V. V. Prokof'eva

Optical instability of the earth's atmosphere impairs the quality of photographs of astronomical objects. One method of improving image quality is to photograph during periods of relative atmospheric quiet.

Two versions of electronic apparatus with an electronic shutter have been created at the Crimean Astrophysical Observatory. They can be used with or without an "atmosphere state indicator." Both devices are based on highly sensitive light receivers [1, 2], which allows them to be used with small telescopes.

A description of the first system — an electronic shutter — was published in [3]. It operates as follows. To "cut off" the image converter, voltage from an independent high-voltage supply equal to the voltage on the next electrode is fed to its photocathode. In this case, there is no image on the image-converter screen. By means of a special externally controlled modulator, the photocathode voltage is reduced to zero or near zero for a specified time interval. The electrode potentials of the image converter become normal, and a brightened image appears on its fluorescent screen. The exposure time is set by a switch in the modulator (within 0.1-0.005 sec). The time lag between the delivery of a command and the beginning of exposure is not over 0.0005 sec, which is entirely sufficient for astronomical purposes. The low weight (3 kg) and small size of the modulator make it suitable for use on medium-size telescopes. It should be noted that this method of image-converter modulation has practically no effect on its resolution.

The second system was described in [4]. It consists of an electronic shutter (whose operation is described above), a television system, and a motion picture camera. Television recording allowed an additional increase in image brightness and an increase in contrast.

A block diagram of the main units is shown in Fig. 1, which also gives diagrams of the voltages in various parts of the circuit. Let us describe one operating cycle. Before the arrival of a control pulse, the electronic shutter 1 and the kinescope 11 are cut off, the image orthicon 4 is in the continuous-readout mode, and the shutter 9 of the motion picture camera is closed. The control pulse I actuates the mechanical shutter of the camera and the electronic shutter. The image from the image-converter screen is projected onto the photocathode of the image orthicon and is stored on its target as a charge pattern. Since pulse IX acts from the moment of delivery of the control pulse, there is no readout during this time. After all of the light has been transferred from the image-converter screen to the image orthicon and stored on its target as a charge pattern and the camera shutter has opened, readout begins (pulse IX, middle notch) and lasts for one television-scanning frame. A coincidence circuit 10 causes this read-

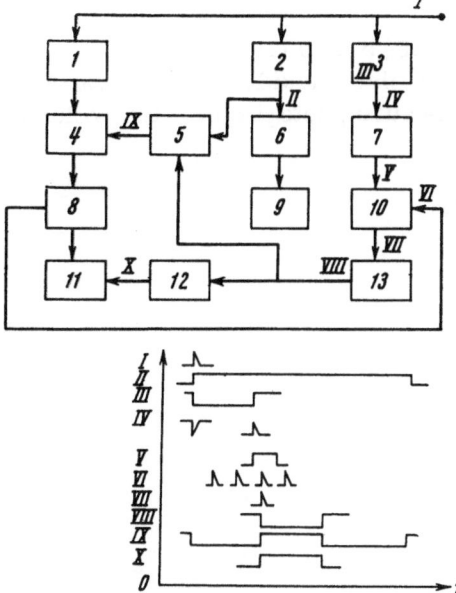

Fig. 1. Block diagram of apparatus and diagram of voltage pulses: 1) electronic shutter; 2) biased multivibrator (τ = 0.4 sec); 3) based multivibrator (τ = 0.1 sec); 4) image orthicon; 5) image-orthicon modulator; 6) èlectronic relay; 7) biased multivibrator (τ = 0.08 sec); 8) television system; 9) shutter of motion picture camera; 10) coincidence circuit; 11) kinescope; 12) kinescope modulator; 13) biased multivibrator (τ = 0.08 sec).

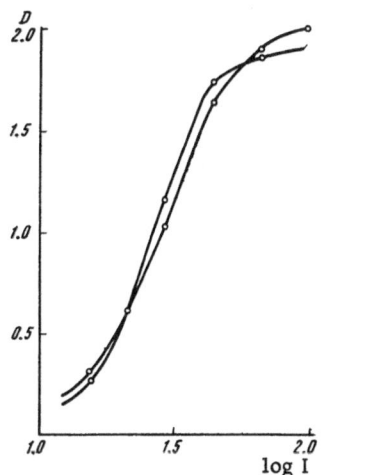

Fig. 2. Characteristics of image converter—television—film system.

out to begin in phase with the frame scanning, which operates continuously. An intensifier pulse X is fed to the kinescope for the entire readout time of the charge pattern, and the image is recorded by the motion picture camera. After the end of pulse II, the camera shutter is closed and the apparatus returns to the initial state.

This system was used in 1965 to check the effectiveness of photographing stars at moments of minimum turbulent atmospheric noise [5]. To standardize the obtained photographs, a device was created for projecting a step-wedge image onto the input photocathode of the system (the photocathode to the image converter). Figure 2 shows curves representing the characteristic of the entire channel (image converter—television system—photographic film). It can be seen that negative blackening is a linear function of the logarithm of the incident-light intensity in a threefold brightness range.

When these devices are used to observe astronomical objects at moments of improved image quality, as determined by some indicator (methods of Platt [6], Gregory [7], Rozhnova [5] et al.), it is desirable to use the same feed optics. This results in some light losses. The guidance requirements of the instrument are determined by the properties of the image-quality indicator.

In short-exposure photography at arbitrary moments (without an image-quality indicator), the probability of obtaining a high-quality photograph is lower [3] and the number of photographs obtained is higher. But the observation time required to obtain one good photograph is only slightly greater, since photographs can be taken more frequently than when an atmosphere-state indicator is used. In some cases, therefore, especially when observing faint stars, and also nebulae, planets, and comets, the described continuous-photographing methods should be used. The control pulses for photographing are generated by an automatic device mounted in the television rack. Advantages of this method are greater penetration and simplification of observations (precise telescope guidance is not necessary).

LITERATURE CITED

1. M. M. Butslov, I. M. Kopylov, et al., Astron. zh., Vol. 39, p. 315 (1962).
2. A. M. Abramenko, A. S. Agapov, et al., Doklady Akad. Nauk SSSR, Vol. 161, No. 6, p. 1299 (1965).

3. A. N. Abramenko, Izv. KrAO, Vol. 34 (1966).
4. A. N. Abramenko and V. V. Prokof'eva, Izv. KrAO, Vol. 35, p. 289 (1966).
5. V. F. Anisimov, R. N. Berdina, N. N. Nechaeva, P. V. Nikolaev, and I. P. Rozhnova, "The probability of obtaining good star images with short-exposure photography," this volume, p. 31.
6. J. R. Platt, Astrophys. J., Vol. 125, p. 601 (1957).
7. R. L. Gregory, Nature, Vol. 203, No. 4942, p. 274 (1964).

ELIMINATING MULTIPLE SCATTERING AND REFLECTION OF LIGHT FROM THE UNDERLYING SURFACE FROM THE SCATTERING INDICATRIX

E. V. Pyaskovskaya-Fesenkova

As is well known, the optical properties of the atmosphere are primarily characterized by two values: optical thickness and light scattering indicatrix. The former value characterizes the total light attenuation in the atmosphere while the latter describes the attenuation due only to scattering and gives information about the luminous fluxes scattered in different directions. While optical thickness can be determined directly, for example, by measuring direct solar radiation, the scattering indicatrix pertains to the entire thickness of the atmosphere, i.e., to a space whose height is equal to the height of the atmosphere and whose basal unit area cannot be determined directly. Brightness observations give only the brightness indicatrix, since additional brightness from higher-order scattering and reflection from the underlying surface is superimposed on the sky brightness caused by atmospheric scattering of sunlight. To obtain the true scattering indicatrix, we must eliminate from the brightness indicatrix this multiply scattered light. This is one of the central problems of atmospheric optics. I proposed a method for this that is apparently simpler than all of the others [1]. Obviously, the error of this method must be known.

The observed brightness of the sky $B(\vartheta)$ at angular distance ϑ (scattering angle) from the sun can be represented as

$$B(\vartheta) = B_1(\vartheta) + B_2(\vartheta), \qquad (1)$$

where $B_1(\vartheta)$ is the brightness due to first-order scattering and $B_2(\vartheta)$ is the additional brightness due to multiple scattering and reflection from the underlying surface.

Let us divide both sides of Eq. (1) by $E_m m$, where E_m is the sun's illumination of a small area perpendicular to the rays at the observation point and m is the atmospheric mass toward the sun. This gives

$$\frac{B(\vartheta)}{E_m m} = \frac{B_1(\vartheta)}{E_m m} + \frac{B_2(\vartheta)}{E_m m}. \qquad (2)$$

Now we shall consider the solar almucantar. From the well-known formula for sky brightness for the solar almucantar caused by only first-order scattering $B_1(\vartheta)$ it follows that the directional-scattering factor

$$\mu_1(\vartheta) = \sigma_1 f_1(\vartheta) = \int_0^\infty \mu_1'(\vartheta)\, dh = \frac{B_1(\vartheta)}{E_m m}, \qquad (3)$$

41

where $\mu_1^!(\vartheta)$ is the directional-scattering factor for an elementary volume of the atmosphere; $f_1(\vartheta)$ is the scattering function or relative scattering indicatrix, which is determined by the properties of the scattering particles and the scattering direction; and σ_1 is a factor determined by the properties and number of scattering particles [1].

In the absence of true absorption, the optical thickness

$$\tau_1 = \tau_p = 2\pi \int_0^\pi \mu_1(\vartheta) \sin\vartheta d\vartheta. \tag{4}$$

But if we consider spectral regions where true absorption occurs, then

$$\tau_1 = \tau_p + \tau_n = 2\pi \int_0^\pi \mu_1(\vartheta) \sin\vartheta d\vartheta + \tau_n, \tag{5}$$

where τ_p and τ_n are the components of optical thickness τ_1 due to scattering and absorption, respectively.

Let

$$\frac{B(\vartheta)}{E_m{}^m} = \mu(\vartheta); \quad \tau = 2\pi \int_0^\pi \mu(\vartheta) \sin\vartheta d\vartheta \tag{6}$$

and call them, respectively, the directional-scattering factor and the optical thickness of the atmosphere that is not free of the effect of multiple scattering and reflection from the underlying surface [1]. In addition, let

$$\frac{B_2(\vartheta)}{E_m{}^m} = \mu_2(\vartheta). \tag{7}$$

This value is the luminous flux scattered at angle ϑ due to multiple scattering and the light reflected from the underlying surface relative to the incident flux, i.e., the additional directional-scattering factor for angle ϑ. Then

$$\mu(\vartheta) = \mu_1(\vartheta) + \mu_2(\vartheta). \tag{8}$$

Assuming that the additional brightness $B_2(\vartheta)$ is independent of azimuth and, therefore, for a given almucantar independent of ϑ, I have proposed a simple method of eliminating these factors from the directional-scattering factor $\mu(\vartheta)$ for various ϑ, solely from observation data, using the formula

$$\mu_1(\vartheta) = \mu(\vartheta) - \frac{\tau - \tau_p}{4\pi}, \tag{9}$$

if true absorption is absent in the spectral region under consideration. But if true absorption is present, formula (9) can be written as

$$\mu_1(\vartheta) = \mu(\vartheta) - \frac{\tau - \tau_1 + \tau_n}{4\pi}, \tag{10}$$

since in this case $\tau_p = \tau_1 - \tau_n$ [1].

In reality, the additional brightness $B_2(\vartheta)$ is to a certain extent dependent upon azimuth. Let us represent $B_2(\vartheta)$ as a sum of two terms one of which is the azimuth-independent component and one of which is azimuth-dependent. Then the additional directional-scattering factor

$$\mu_2(\vartheta) = \frac{\tau - \tau_p}{4\pi} + \mu_2'(\vartheta), \tag{11}$$

where $\mu_2^!(\vartheta)$ can be positive or negative.

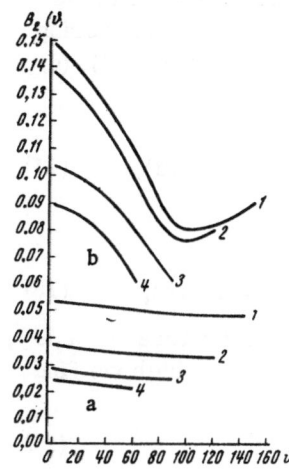

Fig. 1. Additional bright-ness $B_2(\vartheta)$ as a function of scattering angle ϑ: (a) for atmospheric optical thickness $\tau_1 = 0.2$; (b) for $\tau_1 = 0.4$; 1) $z = 75°$; 2) $z = 60°$; 3) $z = 45°$; 4) $z = 30°$.

If we substitute (11) into (8), we have

$$\mu_1(\vartheta) = \mu(\vartheta) - \frac{\tau - \tau_p}{4\pi} - \mu_2'(\vartheta). \tag{12}$$

Let us examine formula (12) for the angle $\vartheta = \vartheta'$ at which the atmospheric scattering indicatrices intersect the spherical for equal τ. As has been shown earlier [1, 2], such an intersection, regardless of the form of the indicatrices, occurs in the atmosphere always near $\vartheta = \vartheta' = 57°$ and when the indicatrices are normalized so that

$$\frac{1}{2}\int_0^{\pi} f(\vartheta)\sin\vartheta\,d\vartheta = 1 \tag{13}$$

always $f(\vartheta') = 1$. This was confirmed later by G. Sh. Livshits [3] and V. E. Pavlov [4]. Therefore,

and also

$$\left.\begin{array}{l} \dfrac{\mu(\vartheta')}{\tau} = \dfrac{1}{4\pi} \quad \text{or} \quad \mu(\vartheta') = \dfrac{\tau}{4\pi}, \\[2ex] \dfrac{\mu_1(\vartheta')}{\tau_1} = \dfrac{1}{4\pi} \quad \text{or} \quad \mu_1(\vartheta') = \dfrac{\tau_1}{4\pi} \end{array}\right\} \tag{14}$$

If we substitute expression (14) into formula (12), we have $\mu_2'(\vartheta') = 0$ and, therefore,

$$\mu_2(\vartheta') = \frac{\tau - \tau_p}{4\pi}. \tag{15}$$

Thus, formula (9) or (10) precisely determines the directional-scattering factor $\mu_1(\vartheta')$ for $\vartheta = \vartheta'$.

For any other scattering angle ϑ, the additional directional-scattering factor $\mu_2(\vartheta)$ is determined by formula (11). If we substitute expression (15) into this formula, we have

$$\mu_2'(\vartheta) = \mu_2(\vartheta) - \mu_2(\vartheta') = \Delta\mu_2(\vartheta). \tag{16}$$

We must find out what the error in $\mu_1(\vartheta)$ for $\vartheta \neq \vartheta'$ will be when formula (9) or (10) is used, in other words, if we let $\mu_2'(\vartheta) = 0$ in expression (12). For this, I used the tables in [5-7], which give $B_2(\vartheta)$, $B(\vartheta)$, and $B(\vartheta):B_1(\vartheta)$ for a number of optical thicknesses τ_1 and solar zenith distances z. Only values for the solar almucantar were taken. The atmosphere is treated as consisting of two layers, the boundary between which is at an altitude with optical thickness $\tau' = {}^3\!/_4\,\tau_1$. In the upper layer the coefficient of asymmetry of the forward- and back-scattered luminous fluxes

$$\Gamma = \frac{\displaystyle\int_0^{\frac{\pi}{2}} \mu_1(\vartheta)\sin\vartheta\,d\vartheta}{\displaystyle\int_{\frac{\pi}{2}}^{\pi} \mu_1(\vartheta)\sin\vartheta\,d\vartheta} = 1.3.$$

In the lower layer, we take several Γ values that correspond to the indicatrices obtained by Foitsik and Chaek in the surface layer.

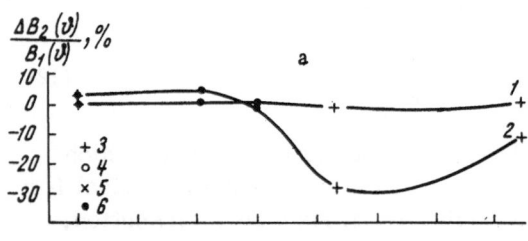

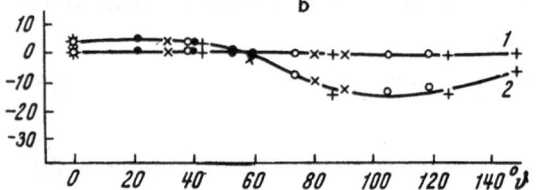

Fig. 2. The difference $\Delta B_2(\vartheta) = B_2(\vartheta) - B_2(57°)$: (a) as a percentage of $B_1(\vartheta)$, the brightness from first-order scattering; (b) as a percentage of $B(\vartheta)$, the brightness from all orders of scattering as a function of the scattering angle ϑ; 1) $\tau_1 = 0.2$; 2) $\tau_1 = 0.4$; 3) $z = 75°$; 4) $z = 60°$; 5) $z = 45°$; 6) $z = 30°$.

Since brightness values are given in these tables, let us convert from directional-scattering factors to brighrnesses. Bearing in mind (3), (6), and (7), we can write formula (12) as

$$B_1(\vartheta) = B(\vartheta) - \frac{\tau - \tau_p}{4\pi} E_m m - B_2'(\vartheta), \qquad (17)$$

expression (16) as

$$B_2'(\vartheta) = B_2(\vartheta) - B_2(\vartheta') = \Delta B_2(\vartheta), \qquad (18)$$

and find the error in $B_1(\vartheta)$ and, therefore, in $\mu_1(\vartheta)$, if we let $\Delta B_2(\vartheta) = 0$ for $\vartheta \neq \vartheta'$, for which we find

$$\frac{\Delta B_2(\vartheta)}{B_1(\vartheta)} = \frac{\Delta \mu_2(\vartheta)}{\mu_1(\vartheta)}. \qquad (19)$$

Since a table with $B_2(\vartheta)$ is given only for $\Gamma = 2.5$ in the lower layer, all of the other values were taken for this same Γ. Figure 1, a and b, shows $B_2(\vartheta)$ as a function of the scattering angle ϑ for the solar almucantar, from [5], for $\tau_1 = 0.2$ and 0.4 and z = 30, 45, 60, and 75°. It is apparent that $B_2(\vartheta)$ for $\tau_1 = 0.2$ varies little with ϑ, and therefore, with azimuth, for all z. For $\tau_1 = 0.4$, the additional brightness $B_2(\vartheta)$ varies considerably for the given Γ.

But formula (17), which gives a brightness values $B_1(\vartheta)$ that is free of the effect of multiple scattering and reflection from the underlying surface, contains not the absolute values of additional brightness $B_2(\vartheta)$ but only their differences $\Delta B_2(\vartheta) = B_2(\vartheta) - B_2(\vartheta')$ for scattering angles ϑ and ϑ'.

Figure 2a gives this difference as a percentage of $B_1(\vartheta)$ for $\tau_1 = 0.2$ (curve 1) and 0.4 (curve 2) as a function of the scattering angle ϑ at solar zenith distances z = 30° (6) and z = 75° (3). With the above-mentioned tables, we could plot $\Delta B_2(\vartheta)/B_1(\vartheta)$ curves with only three points for two z values. For a more complete representation of the error of the method, Fig. 2b shows the difference $\Delta B_2(\vartheta)$ as a percentage of the observed brightness $B(\vartheta)$ for the same τ_1. In this case, it was possible, using the same tables, to calculate $\Delta B_2(\vartheta)/B(\vartheta)$ for z = 30, 45, 60, and 75° (6, 5, 4, 3) for five azimuth (or ϑ) values.

Figure 2 indicates the following about the error of the method due to letting $\Delta B_2(\vartheta) = 0$.

1) The error $\Delta B_2(\vartheta)/B_1(\vartheta)$ is a function of the scattering angle ϑ (or azimuth) and increases slightly with zenith distance z.

2) The maximum error is reached when 85° < ϑ < 120°. The maximum errors $\Delta B_2(\vartheta)/B_1(\vartheta) = \delta$ in % for z = 75° are δ = 1.4 for τ = 0.2 and δ = 28.3 for τ = 0.4.

3) The error for $\tau_1 = 0.2$ is within the error of $B(\vartheta)$ itself for all ϑ.

Thus, it is entirely permissible to ignore the variation in additional brightness $B_2(\vartheta)$ along the solar almucantar for $\tau_1 \leqq 0.2$ when the proposed method is used. The brightness $B_1(\vartheta)$ and the scattering indicatrix $\mu_1(\vartheta)$ are obtained with high accuracy.

Brightness $B_2(\vartheta)$ is a function of the optical thickness τ_1 as well as of the asymmetry of the scattered luminous fluxes Γ. If we consider a "normal" atmosphere, then, according to

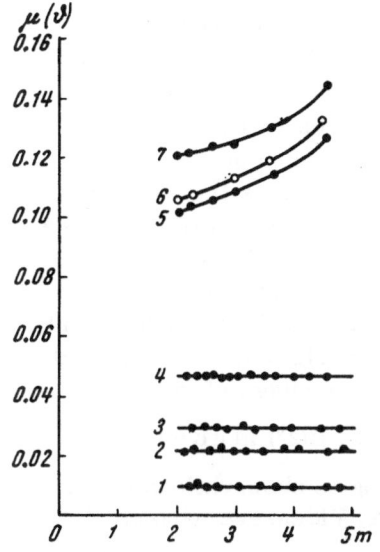

Fig. 3. Directional-scattering factors $\mu(\vartheta)$ aggravated by multiple scattering and reflection from the ground as functions of atmospheric mass m on an optically stable day: 1) $\lambda = 630$ mμ; 2) $\lambda = 495$ mμ; 3) $\lambda = 450$ mμ; 4) $\lambda = 546$ mμ; 6, 7) $\lambda = 333$ mμ; 1, 2, 3, 7) $\vartheta = 60°$; 4) $\vartheta = 2°$ (aureole); 5) $\vartheta = 90°$; 6) $\vartheta = 110°$.

Allen's tables [8], optical thicknesses $\tau_1 = 0.2, 0.4,$ and 0.7 correspond to the following wavelengths:

τ_1	p	λ, mμ
0.2	0.82	540
0.4	0.67	415
0.6	0.55	333

Here, p is the transmission coefficient of the atmosphere. Although τ_1 increases with a decrease in λ, observations have shown that Γ also decreases. According to [1], asymmetry disappears ($\Gamma = 1$) when $\lambda \approx 300$ mμ. For other λ, my observations [9] gave:

λ, mμ	Γ
625	1.60
546	1.54
476	1.37

The effect of increased τ_1 on $B_2(\vartheta)$ is made up for by the decrease in Γ, and the error of the method is therefore small even in the ultraviolet region. To check this, I used Ya. A. Teifel's brightness observations in this spectral region ($\lambda = 333$ mμ).

As early as 1933, V. G. Fesenkov [10] showed theoretically that the relative solar aureole $B(\vartheta)/E_m$, i.e., the ratio of the brightness of the aureole to the sun's illumination of surface perpendicular to the rays, is proportional to the atmospheric mass toward the sun m when the atmosphere is optically stable. Therefore, $B(\vartheta)/E_m$ as a function of m is a straight line that passes through the coordinate origin.

This hypothesis has been confirmed by numerous observations of the brightness of the solar aureole in the visible region by the author and her colleagues of the atmospheric-optics group of the Astrophysical Institute of the Academy of Sciences of the Kazakh SSR. Subsequently, it was shown that a similar relationship holds on the solar almucantar for all ϑ in this spectral region. Thus, in the visible region the effect of multiple scattering is compensated for by other factors that affect brightness.

If we divide $B(\vartheta)/E_m$ for any point on the solar almucantar by the corresponding atmospheric mass m, we obtain a directional-scattering factor $\mu(\vartheta) = B(\vartheta)/E_m m$ that is not free of the effect of multiple scattering or reflection from the underlying surface. For the visible region and an optically stable atmosphere, therefore, $\mu(\vartheta)$ is a constant and is independent of zenith distance. On a graph, $\mu(\vartheta)$ as a function of m will give a straight line parallel to the axis of the abscissas [1]. If any of the factors affecting brightness increases considerably, compensation will no longer occur and the $\mu(\vartheta)$ line will no longer be parallel to the axis of the abscissas.

In fact, in the ultraviolet region, where the effect of multiple scattering is considerable, observations by V. E. Pavlov and later by Ya. A. Teifel' showed that $\mu(\vartheta)$ increased gradually with atmospheric mass m, although the atmosphere had been optically stable during the entire observation time. But as V. E. Pavlov has shown [11], for small scattering angles ϑ (solar aureole) $\mu(\vartheta)$ remains constant within the error limits as m varies, because the brightness in this region is greater than the additional brightness from multiple scattering.

TABLE 1

Curve in Fig. 3	λ, mμ	ϑ	$\dfrac{\mu'(\vartheta)-\mu''(\vartheta)}{\mu'(\vartheta)}$, %	$\dfrac{\mu_1'(\vartheta)-\mu_1''(\vartheta)}{\mu_1'(\vartheta)}$, %
1	630	60	−5	—
2	495	60	0	—
3	450	60	−2.2	—
4	546	2	−4.6	—
5	333	90	20	−4.3
6	333	110	20.6	+2.3
7	333	60	17.5	−2.6

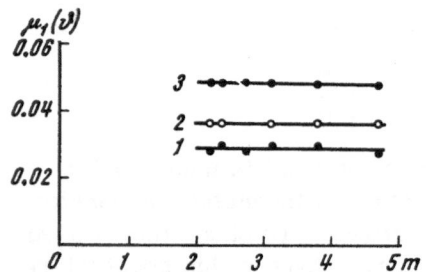

Fig. 4. Directional-scattering factors $\mu_1(\vartheta)$ free of the effect of multiple scattering and ground reflection as functions of atmospheric mass m for λ = 333 mμ: 1) ϑ = 90°; 2) ϑ = 110°; 3) ϑ = 60°.

Figure 3 shows $\mu(\vartheta)$ as a function of m for a number of ϑ and λ values. The observations were made by Ya. A. Teifel' at the mountain observatory of the Astrophysics Institute in the vicinity of Alma-Ata before noon on 4 October 1965. In the table, column 1 gives the number of the lines in Fig. 3, columns 2 and 3 give the wavelengths λ and scattering angles ϑ for which the observations were made, and column 4 gives the difference between $\mu(\vartheta)$ for maximum [$\mu'(\vartheta)$] and minimum [$\mu''(\vartheta)$] atmospheric mass m (first and last observations) as a percentage of $\mu'\vartheta \times \left(\dfrac{\mu'(\vartheta)-\mu''(\vartheta)}{\mu'(\vartheta)}\right)$. In Fig. 3, $\mu(2°)$ is given in an arbitrary scale according to observations made on the same day by T. A. Videneeva, who used a different instrument.

Simultaneous observations at different wavelengths have shown, as is apparent from Fig. 3 and Table 1, that whereas $\mu(\vartheta)$ for various ϑ remains constant within the error limits as atmospheric mass varies in the visible region, for the ultraviolet region $\mu(\vartheta)$ varies rather considerably as m varies from 2 to ~5, owing to the increasing effect of multiple scattering on brightness.

If the proposed method of eliminating multiple scattering and reflection from $\mu(\vartheta)$ can be used in the ultraviolet region despite great optical thickness, then, after applying this method to the data in Fig. 3, $\mu(\vartheta)$ must remain unchanged when m changes. Curves 5, 6, and 7 in Fig. 3, therefore, would have to be straight lines parallel to the axis of the abscissas, as in Fig. 4. Figure 4 gives $\mu_1(\vartheta)$: directional-scattering factors from which the effects of multiple scattering and reflection have been eliminated by the proposed method for λ = 333 mμ and ϑ = 90° (line 1), 110° (line 2), and 60° (line 3). Column 5 of the table gives the difference between $\mu(\vartheta)$ for maximum [$\mu_1'(\vartheta)$] and minimum [$\mu_1''(\vartheta)$] atmospheric mass m as a percentage of $\mu_1'(\vartheta)$.

Summary

1. A method of eliminating the effect of multiple scattering and reflection from the underlying surface from the scattering indicatrix $\mu(\vartheta)$ using only observation data has been proposed. The true scattering indicatrix $\mu_1(\vartheta)$ is given by the formula [1]

$$\mu_1(\vartheta) = \mu(\vartheta) - \frac{\tau - \tau_p}{4\pi}.\qquad(9)$$

This formula was derived on the assumption that the additional brightness B_2 caused by multiple scattering and reflection was independent of azimuth.

2. It has been shown [1, 2] that the atmospheric scattering indicatrices intersect the spherical indicatrix at equal τ always when the scattering angle $\vartheta = \vartheta' \approx 57°$ and always

$f(\vartheta') = 1$ after normalizing the indicatrices to

$$\frac{1}{2}\int_0^\pi f(\vartheta)\sin\vartheta d\vartheta = 1.$$

3. The error of this method due to ignoring the variation of additional brightness $B_2(\vartheta)$ with azimuth in the case of the solar almucantar has been considered. Formula (9) becomes

$$\mu_1(\vartheta) = \mu(\vartheta) - \frac{\tau - \tau_p}{4\pi} - \mu_2'(\vartheta), \tag{12}$$

where $\mu_2'(\vartheta)$ is a term that depends upon the scattering angle ϑ. This formula, therefore, presupposes variability of the additional brightness $B_2(\vartheta)$ with azimuth.

4. It is shown that $\mu_2'(\vartheta) = 0$ for $\vartheta = \vartheta'$ and, hence, that formula (9) is accurate for this scattering angle.

5. On the basis of brightness tables [5-7], it is shown that for $\vartheta \neq \vartheta'$ the error of the method $\Delta\mu_2(\vartheta)/\mu_1(\vartheta)$ depends upon the optical thickness of the atmosphere τ_1, the scattering angle ϑ, and slightly upon the solar zenith distance z. The error of the method is also a function of the asymmetry of the forward- and back-scattered luminous fluxes:

$$\Gamma = \frac{\displaystyle\int_0^{\pi/2} \mu(\vartheta)\sin\vartheta d\vartheta}{\displaystyle\int_{\pi/2}^{\pi} \mu(\vartheta)\sin\vartheta d\vartheta}.$$

A two-layer atmosphere with $\Gamma = 1.3$ in the upper layer and a number of Γ values in the lower is used to compute the tables. The case when $\Gamma = 2.5$ in the lower layer is considered.

6. The error of the method for $\tau_1 \leq 0.2$ is within the error limits for all ϑ. Thus, formula (9) provides sufficient accuracy in this case.

7. The method has its maximum error when $85° < \vartheta < 120°$.

8. As the wavelength λ decreases, the optical thickness of the atmosphere τ_1 increases, but, as observations have shown, Γ decreases in the process. As a result, compensation occurs and the error of the method is small. Observation data show that the method provides high accuracy even in the ultraviolet region.

LITERATURE CITED

1. E. V. Pyaskovskaya-Fesenkova, Investigation of Light Scattering in the Earth's Atmosphere [in Russian], Izd. Akad. Nauk SSSR (1957).
2. E. V. Pyaskovskaya-Fesenkova, Doklady Akad. Nauk SSSR, Vol. 61, No. 6 (1948).
3. G. Sh. Livshits, Light Scattering in the Atmosphere [in Russian], Alma-Ata (1965).
4. E. V. Pavlov, Astron. zh., Vol. 43, No. 4 (1966).
5. E. M. Feigel'son, Izv. Akad. Nauk SSSR, seriya geofiz., No. 10 (1958).
6. E. M. Feigel'son, M. S. Malkevich, S. Ya. Kogan, T. D. Koronatova, K. S. Glazova, and M. A. Kuznetsova, Calculation of the Brightness of Light in the Case of Anisotropic Scattering, Part 1, Consultants Bureau, New York (1960).
7. V. S. Atroshenko, E. M. Feigel'son, K. S. Glazova, and M. S. Malkevich, Calculation of the Brightness of Light in the Case of Anisotropic Scattering, Part 2, Consultants Bureau, New York (1963).
8. C. W. Allen, Astrophysical Quantities, 2nd edition, Oxford University Press, New York (1963).
9. E. V. Pyaskovskaya-Fesenkova, "Light scattering and polarization in the earth's atmosphere," Conference materials, Trudy Astrofiz. Inst. Akad. Nauk Kazakh SSR, Vol. 3 (1962).

10. V. G. Fesenkov, Astron. zh., Vol. 10, No. 3 (1933).
11. V. E. Pavlov, "Study of astroclimate and optical properties of the atmosphere in Kazakh-stan," Trudy Astrofiz. inst. Akad. Nauk Kazakh SSR, Vol. 4 (1963).

SHORT METHOD FOR DETERMINATION OF
ATMOSPHERIC TRANSMISSION COEFFICIENTS

G. A. Kharitonova

Most work in atmospheric optics requires knowledge of the transmission coefficients of the atmosphere in one or another spectral region. Determination of these coefficients by the well-known "long" method of Bouguer entails rather large expenditures of time. Moreover, this method gives accurate values only for an optically stable atmosphere. There is therefore a need for "short" methods.

A number of methods of determining transmission coefficients from the brightness of the daytime clear sky have been proposed by E. V. Pyaskovskaya-Fesenkova [1]. In essence, the methods consist in the following. On the basis of numerous observation data, E. V. Pyaskovskaya-Fesenkova has derived an empirical relationship between the transmission coefficient as determined by the scattering indicatrix and the directional-scattering factor $\mu(\vartheta)$ for four scattering angles ($\vartheta = 40, 60, 90,$ and $120°$), and also between the optical thickness of the atmosphere, again as determined by the scattering indicatrix, and the directional-scattering factor for eight scattering angles. The formulas that express this relationship for $\vartheta = 60, 80,$ and $90°$ are suitable for determining atmospheric transmission coefficients.

For the other angles, the empirical formulas describe only the average dependence of a given directional-scattering factor upon the transmission coefficient. The formula is most accurate for $\vartheta = 60°$. It has the form

$$p_a = 0.973 - 9.8\,\mu\,(60°), \tag{1}$$

where p_a is the transmission coefficient of the atmosphere. The directional-scattering factor $\mu(60°)$ is given by

$$\mu = \frac{A}{\pi} \cdot \frac{B_a}{B_\odot} \cdot \frac{1}{m_\odot}, \tag{2}$$

where A is the albedo of the screen, B_a the brightness at a point on the solar almucantar whose angular distance from the sun $\vartheta = 60°$, $B_\odot$ the brightness of the screen when illuminated by parallel rays of the sun, and $m_\odot$ the atmospheric mass toward the sun.

Transmission coefficients determined by formula (1) are in good agreement with those determined by Bouguer's long method when $p_a \geq 0.82$. At lower p_a, a correction must be made for higher-order scattering.

The aim of the work described here was to derive a formula that could be used over the widest possible range of transmission coefficients without resorting to any corrections. The

49

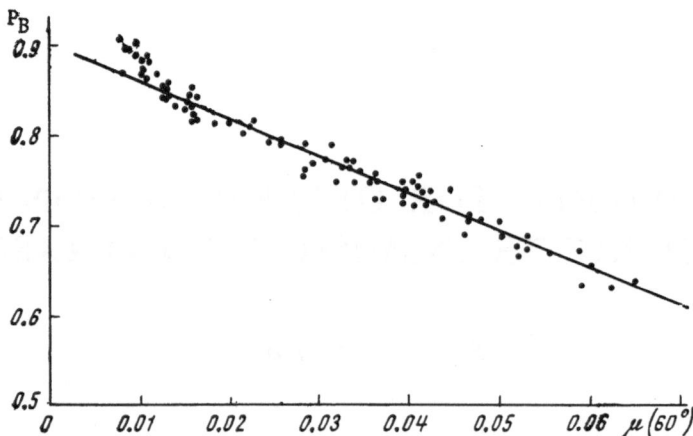

Fig. 1. Relationship between transmission coefficient
p_B and directional-scattering factor $\mu(60°)$.

measurements were made by the author at the observatory of the Astrophysics Institute of the Academy of Sciences of the Kazakh SSR during the summer and autumn of 1965. A portion of the material was obtained in the summer of 1964.

A photoelectric photometer whose detector was an FEU-51 photomultiplier was used. A rotating disk with four interference filters centered at wavelengths λ = 410, 450, 495, and 630 mμ was placed in front of the photomultiplier. After the disk was a diaphragm with a number of circular apertures. The exit angle of the instrument was about 20'. Sensitivity was monitored with the aid of a persistent phosphor, the dependence of the radiation intensity of which upon temperature was the subject of a special study. The phosphor was placed in front of the photomultiplier or slid into a special lightproof cassette, so that stray light could not reach it. The photomultiplier current went through a dc amplifier and was recorded by an EPP-09 M2 potentiometer. A screen was used for measurements of direct solar radiation. It was placed perpendicular to the sun's rays. The photometer was on an altazimuth mounting. Solar angular distances were read with an angular-distance circle.

The measurements were made as follows. First, the brightness of the screen as illuminated by the sun $B_\odot$ was measured. Then the instrument was aimed at the point with ϑ = 60° on the solar almucantar, and the brightness of the sky B_a was measured. After this, the brightness of the screen $B_\odot$ was remeasured. This group of measurements made up a series. As a rule, the phosphor was read after each series. The $B_\odot$ and B_a values were used to determine the directional-scattering factors by formula (2). In addition, after the proper corrections for variation of the sensitivity of the apparatus had been made, the $B_\odot$ values were used to determine the transmission coefficients by the long Bouguer method p_B. It should be pointed out that, as a rule, the sensitivity varied negligibly. Only in rare cases did these variations reach 8-10% during the observations. When the sensitivity variation was ignored, the error in measuring the transmission coefficient in the worst case reached 2-3%. As an example, consider the data obtained on 11 September 1965. On that day, the sensitivity change was 10% as $m_\odot$ varied from 6 to 2. Transmission coefficients obtained using different filters with p_1 and without p_2 taking into account sensitivity change are given below:

λ	p_1	p_2
410	0.63	0.65
450	0.72	0.74
495	0.75	0.77
630	0.83	0.85

Table 1. Comparison of Transmission Coefficients Found by Short p_3 and Long p_B Methods

p_B	p_3	r	p_B	p_3	r	p_B	p_3	r
0.75	0.73	+2.6	0.73	0.74	−1.4	0.79	0.80	−1.3
0.86	0.85	+1.2	0.67	0.65	+3.0	0.86	0.87	−1.2
0.73	0.72	+1.4	0.79	0.80	−1.3	0.76	0.79	−3.9
0.67	0.67	0.0	0.87	0.87	0.0	0.85	0.85	0.0
0.76	0.73	+3.9	0.75	0.77	−2.7	0.71	0.69	+2.8
0.84	0.84	0.0	0.70	0.71	−1.4	0.83	0.84	−1.2
0.77	0.76	+1.3	0.77	0.78	−1.3	0.71	0.72	−1.4
0.88	0.86	+2.2	0.84	0.86	−2.4	0.75	0.76	−1.4
0.75	0.75	0.0	0.73	0.75	−2.7	0.82	0.83	−1.3
0.66	0.68	−3.0	0.68	0.68	0.0	0.68	0.68	0.0
0.75	0.75	0.0	0.76	0.75	+1.3	0.74	0.74	0.0
0.83	0.85	−2.4	0.85	0.84	+1.2	0.82	0.82	0.0
0.72	0.73	−1.4	0.74	0.72	+2.7	0.69	0.68	+1.4
0.63	0.65	−3.1	0.80	0.82	−2.5	0.74	0.73	+1.4
0.75	0.74	+1.3	0.87	0.88	−1.2	0.76	0.76	0.0
0.83	0.84	−1.2	0.76	0.79	−3.9	0.82	0.84	−2.5
0.71	0.71	0.0	0.72	0.73	−1.4	0.87	0.87	0.0
0.64	0.63	+1.6	0.79	0.80	−1.3	0.77	0.78	−1.3
0.73	0.74	−1.4	0.88	0.86	+2.3	0.84	0.85	−1.2
0.81	0.84	−3.7	0.77	0.76	+1.3	0.89	0.87	+2.3
0.69	0.71	−2.9	0.71	0.70	+1.4	0.86	0.86	0.0
0.64	0.64	0.0	0.81	0.82	−1.2	0.91	0.88	+3.3
0.79	0.77	+2.5	0.90	0.88	+2.2	0.77	0.76	+2.6
0.87	0.86	+1.2	0.79	0.79	0.0	0.84	0.84	0.0
0.73	0.75	−2.7	0.74	0.73	+1.4	0.90	0.87	+3.4
0.69	0.69	0.0	0.82	0.81	+1.2	0.78	0.74	+5.2
0.77	0.77	0.0	0.89	0.87	+2.2	0.84	0.84	0.0
0.84	0.85	−1.2	0.74	0.73	+1.4	0.88	0.86	+2.3

The difference between the transmission coefficients for $\lambda = 410\ m\mu$ does not exceed 3%. The difference is less for the other wavelengths. In all, 20 days of observations was used.*

Days on which the directional–scattering factors $\mu(60°)$ had remained constant or varied by not more than 10% during the measurements were selected for determination of the transmission coefficients by the Bouguer method. After calculating $\mu(60°)$ and p_B, we found the mean value of $\mu(60°)$ for the measurement time and plotted the obtained data on a graph (Fig. 1), with p_B on the axis of the ordinates and $\mu(60°)$ on the axis of the abscissas. As can be seen from Fig. 1, the points lie fairly well along a straight line. Then we calculated the correlation coefficient for $\mu(60°)$ and p_B, which turned out to be 0.90. Thus, the relationship between $\mu(60°)$ and p_B is rather close. Well–known formulas were used to derive an expression linking p_B and $\mu(60°)$:

$$p_3 = 0.912 - 4.4\,\mu(60°). \tag{3}$$

We can determine the extent to which the short method agrees with the Bouguer method by examining Table 1, which gives p_3 (the transmission coefficients obtained by formula (3)), those obtained by the long method, and the relative error $r = [(p_B - p_3)/p_B] \cdot 100\%$.

It is apparent that the coefficients calculated by formula (3) agree with those obtained by the long method with accuracy to 1–2%. Sometimes the discrepancy reaches 3–4%. The results

---

*Some of the material was kindly provided by V. E. Pavlov, who made measurements at 450, 542, and 643 mμ using a procedure similar to that used by the author.

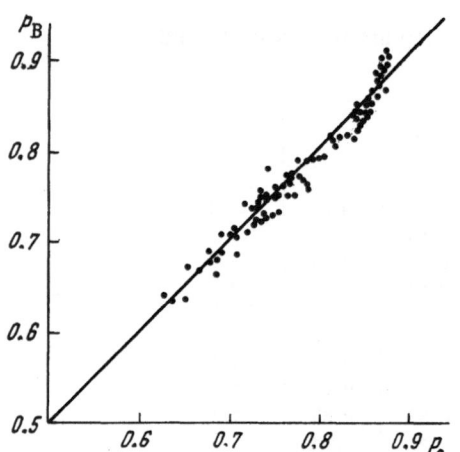

Fig. 2. Comparison of transmission coefficients calculated by short p_3 and long p_B methods.

of the short and long methods are also compared in Fig. 2. This figure shows that for p > 0.85 the points tend to lie on a line that is at an angle to the bisector, i.e., formula (3) does not describe the relationship between p_B and $\mu(60°)$ in this case. Thus, formula (3) can be used with certainty only when $0.60 \leq p \leq 0.85$ or $0.070 \geq \mu(60°) \geq 0.012$. Formula (1) is more suitable for $\mu < 0.012$. Generally speaking, formula (1) is in good agreement with the Bouguer method for $\mu \leq 0.015$, so that formulas (1) and (3) give similar results in the $\mu(60°)$ range from 0.015 to 0.012.

Short-method determination of transmission coefficients according to the brightness of the daytime sky is rather simple and requires little measurement time (1–1.5 min). The method described above will give the transmission coefficient at any given moment, which is particularly important when the atmosphere is optically unstable, when the Bouguer method does not work.

This method can also be used in the absence of snow cover and in the visible region of the spectrum, since formula (3) was derived from observations made under these conditions. Special observations are required to extend this method to other conditions.

I thank Professor E. V. Pyaskovskaya-Fesenkova for formulating the problem and discussing the results, and also V. E. Pavlov for the material he provided.

LITERATURE CITED

1.   E. V. Pyaskovskaya-Fesenkova, Investigation of Light Scattering in the Earth's Atmosphere [in Russian], Izd. Akad. Nauk SSSR, Moscow (1957).

ATMOSPHERIC TRANSMITTANCE AND THE INTERRELATIONSHIPS OF CERTAIN OPTICAL PARAMETERS

G. Sh. Livshits and V. E. Pavlov

Studies of atmospheric transmittance and scattered-light intensity make it possible to establish relationships among certain optical parameters. These relationships can then be used in solving a number of applied problems.

In the present paper, from an analysis of direct observations of spectral transmittance and scattering functions, we shall establish a relationship among the following optical parameters of the atmosphere.

1. Total optical thickness $\tau_B = -\ln p_B$ (p_B is the transmission coefficient as determined by the Bouguer method).

2. The value τ_s, which is determined by numerical integration of the formula

$$\tau_s = 2\pi \int_0^\pi \mu_s(\varphi) \sin \varphi \, d\varphi,$$

where the indicatrix $\mu_s(\varphi)$, which contains the effect of multiple scattering, is found by measuring the brightness of the clear sky on the solar almucantar $B_s(\varphi)$ and the illumination of a perpendicular area at the observation point E. The function $\mu_s(\varphi)$ (φ is the scattering angle, which corresponds to the solar angular distance) is calculated by

$$\mu_s(\varphi) = \frac{B_s(\varphi)}{Em_\odot},$$

where $m_\odot$ is the atmospheric mass toward the sun.

3. The degree of elongation of the scattering function, which is determined by

$$\Gamma = \frac{\int_0^{\pi/2} \mu_s(\varphi) \sin \varphi \, d\varphi}{\int_{\pi/2}^\pi \mu_s(\varphi) \sin \varphi \, d\varphi}.$$

4. The optical thickness for pure absorption τ_n.

53

The interrelationship of these parameters was found on the basis of the following considerations. Both observations and theoretical data indicate a very slight angular dependence of brightness component caused by multiple scattering (B_2) and reflection from the underlying surface (B_q). The values of $B_2 + B_q$ increase somewhat at very small ($\varphi \approx 0°$) and very large ($\varphi \approx 180°$) scattering angles. Therefore, the fluxes

$$\Phi_2 = 2\pi \int_0^\pi \frac{B_2(\varphi)}{Em_\odot} \sin\varphi d\varphi \quad \text{and} \quad \Phi_q = 2\pi \int_0^\pi \frac{B_q(\varphi)}{Em_\odot} \sin\varphi d\varphi,$$

due, respectively, to multiple scattering and reflection from the underlying surface, are distributed approximately equally in the front and back hemispheres (as compared with direct rays):

$$\Phi_{2(0°)} = 2\pi \int_0^{\pi/2} \frac{B_2(\varphi)}{Em_\odot} \sin\varphi d\varphi \approx \Phi_{2(180°)} = 2\pi \int_{\pi/2}^\pi \frac{B_2(\varphi)}{Em_\odot} \sin\varphi d\varphi,$$

$$\Phi_{q(0°)} = 2\pi \int_0^\pi \frac{B_a(\varphi)}{Em_\odot} \sin\varphi d\varphi \approx \Phi_{q(180°)} = 2\pi \int_0^\pi \frac{B_q(\varphi)}{Em_\odot} \sin\varphi d\varphi.$$

On the other hand, studies of the aerosol scattering function $\mu_a(\varphi)$ from observations of daytime clear-sky brightness show that the fluxes scattered forward by aerosol are much greater than those scattered back:

$$2\pi \int_0^{\pi/2} \mu_a(\varphi) \sin\varphi d\varphi \gg 2\pi \int_{\pi/2}^\pi \mu_a(\varphi) \sin\varphi d\varphi.$$

In particular, attempts to determine $\mu_a(\varphi)$ for $\varphi > 90°$ have not met with success, since the found values are within the instrument error.

We can use these facts to solve a number of problems with sufficient accuracy, proceeding from two simplifying assumptions: the multiply scattered luminous fluxes are symmetric relative to scattering angle $\varphi = 90°$ and the luminous flux scattered back by aerosol is negligible.

These assumptions make it easy to establish an interrelationship among the above-listed parameters as follows:

$$\tau_B = \tau_R + \tau_s \left(\frac{\Gamma-1}{\Gamma+1}\right) + \tau_n, \tag{1}$$

where τ_R is the Rayleigh optical thickness.

Formula (1) is trivial if the aerosol optical thickness is determined by

$$\tau_a = \tau_s \left(\frac{\Gamma-1}{\Gamma+1}\right). \tag{2}$$

Equation (2) is obtained from elementary considerations, if we start from the above assumptions. In fact, if we assume that when $\varphi > 90°$ the scattered-light intensity is determined only by Rayleigh and multiple scattering, subtracting from τ_s twice the value of the integral

$$\tau_{s(180°)} = 2\pi \int_{\pi/2}^\pi \mu_s(\varphi) \sin\varphi d\varphi$$

will give the aerosol optical thickness

$$\tau_a = \tau_s - 2\tau_{s(180°)}. \tag{3}$$

If we incorporate into this formula the asymmetry $\Gamma = \tau_{s(0°)}/\tau_{s(180°)}$, we obtain formula (2).

Of course, formulas (1) and (2) are valid within specific limits and can only be used in solving problems that do not require greater than 2-3% accuracy for τ.

The approximate nature of formula (1) is a result of the simplifications made above. If we assume that the term $B_2 + B_q$ is symmetric relative to $\varphi = 90°$ and subtract double the value of the back-scattered flux to determine τ_a, we somewhat overestimate this value, because the indicatrix effect increases somewhat the role of multiple scattering at small φ. On the other hand, if we assume that aerosol causes entirely no back scattering, we definitely decrease τ_a. Determination of the extent to which the errors balance each other out in each specific case is a rather complicated task. Therefore, the formula (1) should be evaluated on the basis of direct observations after it has been used to solve a particular problem. In particular, formula (1) was tested as the basis of a method of determining transmission coefficients.

Measurements of daylight clear-sky brightness $B_s(\varphi)$ made by a number of observers under various conditions were used for this. The transmission coefficients p_B were determined by the Bouguer method, and in this way the left ($\tau_B = -\ln p_B$) as well as the right sides of (1) could be found independently. We set τ_n according to the mean ozone content (0.3 cm). The results of these studies led us to the conclusion that formula (1) could serve as the basis of a short method of determining p_B that would have great advantages. It can be used over a wide range of optical thicknesses, wavelengths, and observation conditions, in summer as well as in winter, and in the presence of snow cover.

The difference between τ_B as determined by the Bouguer method and by formula (1) are somewhat greater in the longwave part of the visible region than in the shortwave part. This may be explained by three factors: 1) the role of aerosol is more considerable at higher λ; 2) the symmetry of the intensity of secondarily scattered light decreases as the role of Rayleigh scattering decreases; and 3) the optical thicknesses for ozone absorption must be given with greater accuracy (we used average values). Nevertheless, for many purposes, if the accuracy satisfies the requirements of the problem, the formula can also be used in the longwave part of the visible region.

The parameter interrelationship described by formula (1) opens up the possibility of solution of other interesting problems. For example, one problem of ozonometry is determination of the O_3 content of the atmosphere with allowance made for aerosol extinction of light.

If aerosol extinction in the Hartley band is fairly great and, in addition, varies considerably over the spectrum, serious difficulties will be encountered in determining the ozone content, because the components of the optical thicknesses that are to be determined by the Bouguer method must be separated.

If we use formula (1), which is most accurate in the ultraviolet region, we can isolate the desired optical thickness for pure absorption according to the difference between the observed and calculated τ_B values. Aerosol scattering can be calculated on the basis of formula (2). It follows from our results that in the near-ultraviolet region, outside the Hartley band, the role of pure absorption in aerosols is small. Otherwise, there would be no agreement between the observed and calculated τ_B. This provides a basis for assuming that in ozone determination in a rural, mountainous area, we can isolate the component of aerosol optical thickness that is caused solely by scattering. This simplifies the matter considerably, since formula (1) permits solution of the ozonometric problem with allowance for aerosol extinction:

$$\tau_n = \tau_B - \tau_s \left(\frac{\Gamma - 1}{\Gamma + 1}\right) - \tau_R.$$

Aerosol extinction due to scattering is automatically taken into account here.

Thus, solution of the problem becomes comparatively simple if, when determining transmittance by the standard Bouguer method, we simultaneously measure the brightness on the solar almucantar. It should be added that calculation of τ_s does not require data on the directional-scattering factors $\mu_s(\varphi)$ for $\varphi > 90°$, because in this part of the sky the brightness is almost entirely determined by Rayleigh scattering. This fact can also be used in determining the albedo of a locality from sky brightness. After having determined p_B and the observed brightness $B_s(\varphi)$ at $\varphi > 90°$ and subtracted from these values $B_n(\varphi)$, the components of Rayleigh $B_R(\varphi)$ and multiple scattering $B_2(\varphi)$, we find the component $B_q(\varphi)$, which is due to atmospheric scattering of reflected radiation. The latter component is a function of the albedo q, so knowledge of $B_q(\varphi)$ makes it possible to find the parameter q. This problem can be solved when atmospheric transmittance is fairly high, since at $\varphi > 90°$ the intensity is determined almost entirely by Rayleigh scattering.

MONITORING THE OPTICAL STABILITY OF THE ATMOSPHERE FOR SMALL AND LARGE OPTICAL THICKNESSES

V. E. Pavlov

The brightness of the sky on the solar almucantar is chiefly determined by the following parameters [1, 2]: the solar zenith distance $z_\odot$, the optical thickness of the atmosphere τ_1, and the albedo of the underlying surface q, and it is greatly dependent upon the form of the scattering indicatrix.

The sphericity of the earth can be fairly reliably taken into account by replacing the secants of the solar zenith distances by the corresponding atmospheric masses [1, 2]. After this substitution, the observed brightness of the sky can be represented as a sum of three terms:

$$B(\vartheta, m, \tau_1, q) = B_1(\vartheta, m, \tau_1) + B_2(\vartheta, m, \tau_1) + B_q(\vartheta, m, \tau_1, q), \tag{1}$$

where B_1 is the single-scattering brightness, B_2 the multiple scattering brightness, B_q the brightness due to reflection from the underlying surface with subsequent scattering in the atmosphere, ϑ the scattering angle, and m the atmospheric mass toward the sun.

When using relation (1), bear in mind the following three facts: we shall consider only the stationary problem, i.e., the optical properties of the atmosphere remain constant during observations; the atmosphere is assumed to be horizontally homogeneous; and it is assumed to be a purely scattering medium.

If we divide the left and right sides of (1) by $E_0 p^m m$, where E_0 is the solar constant and p is the Bouguer atmospheric transmission coefficient, and let

$$\mu(\vartheta, m, q) = \frac{B(\vartheta, \tau_1, m, q)}{E_0 p^m m}$$

and

$$\mu_1(\vartheta, m) = \frac{B_1(\vartheta, \tau_1, m)}{E_0 p^m m},$$

we obtain

$$\mu(\vartheta, m, q) = \mu_1(\vartheta, m) + \frac{B_2(\vartheta, \tau_1, m)}{E_0 p^m m} + \frac{B_q(\vartheta, \tau_1, m, q)}{E_0 p^m m}, \tag{2}$$

where $\mu_1(\vartheta, m)$ is the weighted-mean scattering indicatrix for the entire height of the atmosphere.

Theoretical calculations show [2, 3] that for points on the solar almucantar the averaging weights are independent of the solar zenith distance, i.e., μ_1 is not a function of atmospheric mass.

Light reflection from the underlying surface is usually considered orthotropic. Attempts to determine the effect of an anisotropic underlying surface (snow cover, in particular) on brightness have not given significant results [2]. In view of this, we shall assume that B_q is independent of the scattering angle ϑ, although in general the absolute value of B_q may vary somewhat with the form of the scattering indicatrix [4].

Considering the above, Eq. (2) can be rewritten as

$$\mu(\vartheta, m, q) = \mu_1(\vartheta) + \frac{B_2(\vartheta, m, \tau_1)}{E_0 p^m m} + \frac{B_q(m, \tau_1, q)}{E_0 p^m m}. \tag{3}$$

In the case of pure scattering, the optical thickness

$$\tau_1 = 2\pi \int_0^\pi \mu_1(\vartheta) \sin \vartheta d\vartheta. \tag{4}$$

Constancy of the absolute indicatrix μ_1 is a good criterion of time stability of the scattering properties of the earth's atmosphere. But only $\mu(\vartheta, m, q)$ (the absolute brightness indicatrix) or the scattering indicatrix containing the effects of multiple scattering and reflection from the underlying surface can be obtained directly from brightness observations [1]. At small scattering angles, owing to elongation of the indicatrix in the direction of solar-ray incidence, the role of the second and third terms on the right of formula (3) is small, and the observed brightness indicatrix practically coincides with the scattering indicatrix if the optical thickness is not too great. This underlies the stability criterion for the optical properties of the earth's atmosphere developed by V. G. Fesenkov [5] and later added to and refined by E. V. Pyaskovskaya-Fesenkova [1]. In fact, constancy of the relative solar aureole (the ratio of the illuminations from the aureole and from the sun) per unit atmospheric mass indicates constancy of the scattering properties of the earth's atmosphere.

As the scattering angle increases, the relative contributions of the second and third terms on the right of (3) also increase. Nevertheless, brightness observations in the visible region [1] indicate that when $\mu(\vartheta)$ is constant for the aureole it is also constant for any other specified scattering angle.

We must resort to theoretical calculations of brightness in order to evaluate the contributions of each of the components $B_2/E_0 p^m m$ and $B_q/E_0 p^m m$ to the observed brightness indicatrix $\mu(\vartheta)$ at various solar zenith distances. For this we used tables calculated assuming anisotropic scattering in a two-layer atmosphere with a layerwise varying scattering indicatrix [4]. A property of atmospheric indicatrices discovered by E. V. Pyaskovskaya-Fesenkova [1] was used to interpret the brightness observations at large scattering angles. When

$$\tau = 2\pi \int_0^\pi \mu(\vartheta) \sin \vartheta d\vartheta \tag{5}$$

the absolute brightness indicatrices intersect the spherical indicatrix near $\vartheta = 57°$. A similar situation occurs for the initial scattering act: the absolute scattering indicatrices $\mu_1(\vartheta)$ also intersect near $\vartheta = 57°$ when the optical thicknesses are equal (see (4)).

The values $\mu(57°)$ and τ and $\mu_1(57°)$ and τ_1 are related:

$$\tau = 4\pi\mu(57°) \tag{6}$$

and

$$\tau_1 = 4\pi\mu_1(57°).\tag{7}$$

These relations and the brightness tables permit quantitative calculation of each of the components in Eq. (3). It turned out that the theoretical calculations were in fairly good agreement with the observation data: within the measurement error $F(0°)/F_{\odot}m$ and $F(57°)/F_{\odot}m$ were independent of solar zenith distance.

Let us study the diurnal variation of each of the components of (3). If $\tau_1 = 0.2$, then $\mu_1(57°) = 0.01592$, wherein, according to the above, this value is independent of the solar zenith distance. The values of $B_2(57°)/E_0p^m m$ and $B_q/E_0p^m m$ as functions of m are:

$\overset{\bullet}{z}_{\odot}$	m	$\mu_1(57°)$	$\dfrac{B_2(57°)}{E_0 p^m m}$	$\dfrac{B_q}{E_0 p^m m}$	$\mu(57°)$
30	1.154	0.01592	0.00381	0.00431	0.0240
45	1.413	0.01592	0.00383	0.00365	0.0234
60	2.000	0.01592	0.00415	0.00324	0.0233
75	3.816	0.01592	0.00458	0.00172	0.0222

It can be seen that $B_2(57°)/E_0p^m m$ increases with an increase in m but $B_q/E_0p^m m$ decreases. In the final analysis, compensation occurs such that the deviations from the mean value of $\mu(57°)$ do not exceed 4.5% (the error in determining $\mu(\vartheta)$ from observations is usually 4–5%).

Let us consider the dependence of $\mu(\vartheta)$ on $z_{\odot}$ for large optical thicknesses. For this we shall use brightness observations in the ultraviolet region ($\lambda = 347.5$ mμ). The preliminary results were discussed earlier [7].

It was shown that $F(2°15')/F_{\odot}m$ is retained over the entire range of atmospheric masses $m \leq 6$, while $F(100°)/F_{\odot}m$ is retained only at $m \leq 3$. Starting with $m > 3$, $F(100°)/F_{\odot}m$ increases with an increase in atmospheric mass. This latter effect occurs over the entire range of large scattering angles ($20° \leq \vartheta \leq 160°$), with the exception of the solar aureole [7].

Just as in the preceding case, we shall use theoretical brightness calculations to interpret the observation data. First of all, we shall show that relations (6) and (7), which make it rather easy to separate the observed brightness into its components, remain valid in the ultraviolet region.

The question of intersection of the indicatrices $\mu_1(\vartheta)$ near $\vartheta = 57°$ in the case of large optical thicknesses must be given special consideration. For example, O. D. Barteneva has shown [8] that in the atmospheric boundary layer the point of intersection of the indicatrices is shifted in the direction of small scattering angles as optical thickness increases. It should be noted that as optical thickness in the atmospheric boundary layer increases, the elongation of the indicatrix in the direction of incidence of the rays illuminating the scattering body also increases. If the form of the indicatrix is characterized by the ratio of the forward- and back-scattered luminous fluxes

$$\Gamma_1 = \frac{\int_0^{\pi/2} \mu_1(\vartheta)\sin\vartheta\,d\vartheta}{\int_{\pi/2}^{\pi} \mu_1(\vartheta)\sin\vartheta\,d\vartheta},\tag{8}$$

then in the atmospheric boundary layer Γ_1 will increase with an increase in τ_1. In the presence of high turbidity, usually $\Gamma_1 > 3$.

We can estimate Γ_1 for the entire atmosphere in the ultraviolet region as follows. Let us represent the optical thickness τ_1 as the sum of its Rayleigh and aerosol components:

$$\tau_1 = \tau_R + \tau_a. \tag{9}$$

We shall assume that aerosols scatter light only forward, in the direction of solar-ray incidence. In this case,

$$\Gamma_1 = \frac{\tau_a + \frac{1}{2}\tau_R}{\frac{1}{2}\tau_R}. \tag{10}$$

Calculation for the territory of the observatory of the Academy of Sciences of the Kazakh SSR, where the brightness observations were made, gives $\tau_R = 0.553$. At the observation point, τ_1 usually varies from 0.6 to 0.8. In this case, the asymmetry parameter for the scattered luminous fluxes Γ_1 from formula (10) will not exceed 1.18-1.91. At such low values of Γ_1, for equal optical thicknesses the scattering indicatrices will always intersect near $\vartheta = 57°$ in the atmospheric boundary layer [8] as well as throughout the entire atmosphere [1, 2].

Thus, relation (7) must hold not only in the visible but also in the ultraviolet region.

A detailed analysis of observation data can be made by using relations (6) and (7) and the appropriate brightness tables. Investigation of the form of the indicatrix in the ultraviolet region [9] indicates that the elongated scattering indicatrix must be used in calculations for the region of the solar aureole (see, for example, the tables in [4]). In the area of $\vartheta = 57°$, calculations can be made with the spherical indicatrix, since near $\vartheta = 57°$ the actual and spherical indicatrices intersect at equal optical thicknesses. The tables of E. S. Kuznetsov and B. V. Ovchinskii [10] were used in our calculations.

The value $F(0°)/F_\odot m$ is practically independent of m, while $F(57°)/F_\odot m$ increases with an increase in m.

The increase in $F(57°)/F_\odot m$ with m may be explained by either the second or third term on the right side of (3). Calculation shows that $B_q/E_0 p^m m$ decreases with an increase in solar zenith distance, and the contribution of this component is small when $q = 0.1-0.2$. The observed effect, therefore, is due to variation of the contribution of $B_2/E_0 p^m m$. As the sun approaches the horizon, in the ultraviolet region the solar aureole seems to be blurred against the background of the veil created by multiply scattered light.

Thus, the stability of the optical properties of the atmosphere for great optical thicknesses can be monitored only according to the variation of the relative solar aureole. The range of large scattering angles is unsuitable for this purpose.

The fact that $\mu(\vartheta)$ increases with an increase in m for great optical thicknesses also limits the applicability of E. V. Pyaskovskaya-Fesenkova's short method for determining atmospheric transmission coefficients from brightness measurements at a point with $\vartheta = 60°$ [1], since her method is based on the condition that $\mu(60°)$ be independent of solar zenith distance.

<h2 style="text-align:center">LITERATURE CITED</h2>

1. E. V. Pyaskovskaya-Fesenkova, Investigation of Light Scattering in the Earth's Atmosphere [in Russian], Izd. Akad. Nauk SSSR, Moscow (1957).
2. G. Sh. Livshits, "Light scattering in the atmosphere," Trudy Astrofiz. inst. Akad. Nauk Kazakh SSR, Vol. 6 (1965).
3. E. M. Feigel'son, "Interpreting sky brightness observations," Izv. Akad. Nauk SSSR, seriya geofiz., No. 10, p. 1222 (1958).

4. E. M. Feigel'son, M. S. Malkevich, S. Ya. Kogan, T. D. Koronatova, K. S. Glazova, and M. A. Kuznetsova, Calculation of the Brightness of Light in the Case of Anisotropic Scattering, Part 1, Consultants Bureau, New York (1960).

5. V. G. Fesenkov, "On the problem of determining the solar constant," Astron. zh., Vol. 10, No. 3, p. 249 (1933).

6. V. E. Pavlov, "A daylight photoelectric recording photometer and the light-scattering indicatrix in the earth's atmosphere, including small scattering angles," in: Light Scattering and Polarization in the Earth's Atmosphere, Alma-Ata (1962), p. 62.

7. V. E. Pavlov, "The light-scattering indicatrix in the earth's atmosphere in the ultraviolet region," in: Studies of Astroclimate and Optical Properties of the Atmosphere in Kazakhstan, Alma-Ata (1963), p. 93.

8. O. D. Barteneva, "Scattering indicatrices in the atmospheric boundary layer," Izv. Akad. Nauk SSSR, seriya geofiz., No. 12, p. 1853 (1960).

9. V. E. Pavlov, "The atmospheric scattering indicatrix in the visible and ultraviolet regions," Astron. zh., Vol. 41, No. 3, p. 546 (1964).

10. E. S. Kuznetsov and B. V. Ovchinskii, "Results of numerical solution of the integral equation of the theory of atmospheric light scattering," Trudy Geofiz. inst., No. 4, p. 131 (1949).

SPECTRAL BRIGHTNESS OF THE SKY

A. I. Ivanov

Knowledge of the brightness distribution of the daytime clear sky for different values of the atmospheric parameters is necessary for solving a number of applied problems. In twilight studies, for example, in order to take into account the influence of the troposphere, the brightness of the daytime clear sky must be known in absolute units for a large set of parameters. Experimental data on the brightness distribution make it possible to judge how close a particular model of the atmosphere used in theoretical calculations comes to reality. So far, however, there are no comprehensive tables of the spectral brightness of the daytime clear sky for a wide range of parameters, or even for the most important parameter — transmittance. The available results, for example, [1-3], were obtained only for a small set of parameters using visual and photoelectric photometers with wideband filters. In a number of cases, transmittance was not determined [2], and the results were given in relative units (readings from various points in the sky were tied to the zenith).

We made measurements of the spectral brightness of the sky with an automatic-recording photoelectric spectrophotometer. The instrument was based on a UM-2 monochromator, the radiation detector was an FEU-51 photomultiplier, and an EPP-09M electronic potentiometer served as the recorder. A gypsum screen illuminated by direct solar radiation was used for standardization. The instrument and the operating procedure when a loop oscillograph is used as the recorder have been described in detail by P. N. Boiko [7].

Measurements were made over the entire sky with azimuth and vertical-circle steps $\Delta\Psi = 45°$ and $\Delta z = 15°$, respectively, at wavelengths $\lambda = 6910, 6500, 5930, 5530, 4470,$ and 4040 Å at a number of solar zenith distances.

During the measurements, the optical stability of the atmosphere was monitored by the method proposed in [1]. The atmosphere was considered optically stable if the directional-scattering factor for the entire height of the atmosphere, which we determined for a point on the solar almucantar with angular distance $\theta = 20°$ from the sun, remained constant within the measurement error for the entire observation period. We processed only the observations for optically stable days and days when transmittance worsened or improved gradually. In the latter case, on the basis of [3], the transmittance may be defined as the arithmetic mean of the transmittances obtained by the Bouguer method (p_B) and by the formula

$$\ln \frac{B_s}{m_\odot} = m_\odot \ln p_s + C, \tag{1}$$

TABLE 1. Spectral Brightness 100B (W/cm² · sr · μ)

27 August 1965, A. M.

λ = 0.691; p = 0.86±0.01; τ = 0.15

z°	z_⊙ = 78.1°					z_⊙ = 70.4°					z_⊙ = 63.3°				
	0	45	90	135	180	0	45	90	135	180	0	45	90	135	180
0	0.064	0.064	0.064	0.064	0.064	0.083	0.083	0.083	0.083	0.083	0.112	0.112	0.112	0.112	0.112
15	0.090	0.080	0.064	0.056	0.055	0.120	0.105	0.081	0.066	0.063	0.168	0.141	0.107	0.087	0.084
30	0.140	0.107	0.068	0.055	0.055	0.220	0.105	0.086	0.067	0.064	0.321	0.205	0.112	0.082	0.077
45	0.320	0.166	0.083	0.069	0.072	0.550	0.235	0.100	0.079	0.082	—	0.301	0.127	0.091	0.092
60	—	0.290	0.112	0.099	0.108	—	0.366	0.140	0.114	0.123	—	0.446	0.169	0.126	0.132
75	—	0.540	0.208	0.203	0.215	—	0.549	0.239	0.213	0.239	—	0.658	0.267	0.231	0.249

λ = 0.553; p = 0.82±0.01; τ = 0.20

z°	z_⊙ = 76°					z_⊙ = 68.6°					z_⊙ = 61.3°				
	0	45	90	135	180	0	45	90	135	180	0	45	90	135	180
0	0.149	0.149	0.149	0.149	0.149	0.198	0.198	0.198	0.198	0.198	0.248	0.248	0.248	0.248	0.248
15	0.204	0.184	0.153	0.139	0.137	0.281	0.247	0.196	0.169	0.163	0.361	0.313	0.245	0.205	0.195
30	0.320	0.244	0.186	0.144	0.146	0.481	0.345	0.217	0.175	0.170	0.693	0.440	0.260	0.202	0.190
45	0.643	0.356	0.201	0.182	0.192	1.04	0.502	0.247	0.208	0.211	—	0.641	0.309	0.234	0.226
60	—	0.587	0.280	0.266	0.302	—	0.755	0.335	0.297	0.327	—	0.887	0.394	0.322	0.340
75	—	1.01	0.514	0.494	0.570	—	1.19	0.554	0.541	0.614	—	1.30	0.693	0.575	0.635

λ = 0.404; p = 0.66±0.01; τ = 0.42

z°	z_⊙ = 72.5°					z_⊙ = 65.8°					z_⊙ = 58.7°				
	0	45	90	135	180	0	45	90	135	180	0	45	90	135	180
0	0.332	0.332	0.332	0.332	0.332	0.438	0.438	0.438	0.438	0.438	0.546	0.546	0.546	0.546	0.546
15	0.406	0.378	0.331	0.310	0.303	0.579	0.524	0.443	0.402	0.398	0.734	0.660	0.546	0.485	0.462
30	0.610	0.495	0.381	0.353	0.353	0.848	0.669	0.481	0.417	0.408	1.13	0.842	0.582	0.492	0.464
45	1.04	0.681	0.449	0.445	0.463	1.528	0.907	0.558	0.506	0.516	—	1.10	0.656	0.568	0.568
60	—	0.951	0.575	0.606	0.669	—	1.19	0.699	0.688	0.741	—	1.41	0.801	0.747	0.788
75	—	1.28	0.794	0.878	0.993	—	1.56	0.950	1.01	1.12	—	1.71	1.08	1.10	1.20

A. I. IVANOV

Table 1 (cont.)

4 September 1965, A. M.

z_⊙ = 76.6°

$z°$	0	45	90	135	180
0	0.050	0.050	0.050	0.050	0.050
15	0.067	0.062	0.051	0.049	0.044
30	0.118	0.087	0.057	0.049	0.049
45	0.262	0.131	0.067	0.058	0.063
60	—	0.218	0.094	0.090	0.101
75	—	0.418	0.170	0.174	0.204

z_⊙ = 74.7°

$z°$	0	45	90	135	180
0	0.066	0.066	0.066	0.066	0.066
15	0.092	0.082	0.066	0.060	0.058
30	0.153	0.114	0.074	0.061	0.062
45	0.336	0.167	0.089	0.076	0.082
60	—	0.286	0.124	0.117	0.133
75	—	0.509	0.223	0.230	0.269

z_⊙ = 72.4°

$z°$	0	45	90	135	180
0	0.286	0.286	0.286	0.286	0.286
15	0.364	0.332	0.289	0.270	0.264
30	0.532	0.436	0.322	0.295	0.300
45	0.888	0.581	0.386	0.363	0.395
60	—	0.855	0.523	0.537	0.612
75	—	1.31	0.791	0.889	1.05

$\lambda = 0.691;\; p = 0.89 \pm 0.01;\; \tau = 0.11$

z_⊙ = 69.6°

$z°$	0	45	90	135	180
0	0.061	0.061	0.061	0.061	0.061
15	0.091	0.080	0.062	0.054	0.052
30	0.163	0.110	0.066	0.054	0.052
45	0.394	0.167	0.079	0.064	0.066
60	—	0.265	0.109	0.095	0.104
75	—	0.453	0.191	0.188	0.216

$\lambda = 0.650;\; p = 0.88 \pm 0.01;\; \tau = 0.13$

z_⊙ = 67.7°

$z°$	0	45	90	135	180
0	0.080	0.080	0.080	0.080	0.080
15	0.114	0.100	0.079	0.068	0.065
30	0.207	0.142	0.086	0.068	0.068
45	0.495	0.204	0.103	0.084	0.085
60	—	0.328	0.140	0.126	0.138
75	—	0.550	0.251	0.251	0.283

$\lambda = 0.447;\; p = 0.74 \pm 0.01;\; \tau = 0.30$

z_⊙ = 65.7°

$z°$	0	45	90	135	180
0	0.340	0.340	0.340	0.340	0.340
15	0.443	0.402	0.344	0.311	0.301
30	0.663	0.524	0.376	0.318	0.318
45	1.22	0.710	0.444	0.400	0.419
60	—	1.01	0.598	0.579	0.640
75	—	1.49	0.900	0.976	1.12

z_⊙ = 60°

$z°$	0	45	90	135	180
0	0.080	0.080	0.080	0.080	0.080
15	0.125	0.105	0.078	0.064	0.060
30	0.248	0.149	0.083	0.059	0.057
45	—	0.218	0.092	0.069	0.069
60	—	0.316	0.124	0.101	0.105
75	—	0.456	0.207	0.199	0.223

z_⊙ = 58.4°

$z°$	0	45	90	135	180
0	0.103	0.103	0.103	0.103	0.103
15	0.158	0.134	0.101	0.084	0.078
30	0.333	0.189	0.106	0.081	0.074
45	—	0.264	0.119	0.090	0.088
60	—	0.366	0.157	0.130	0.138
75	—	0.552	0.267	0.258	0.284

z_⊙ = 56.1°

$z°$	0	45	90	135	180
0	0.441	0.441	0.441	0.441	0.441
15	0.612	0.539	0.444	0.388	0.375
30	0.958	0.703	0.482	0.388	0.375
45	—	0.920	0.542	0.458	0.463
60	—	1.20	0.700	0.639	0.681
75	—	1.63	1.36	1.08	1.21

Table (cont.)

6 September 1965, A. M.

λ = 0.691; p = 0.84 ± 0.01; τ = 0.18

z°	z_⊙ = 75.6°					z_⊙ = 68.9°					z_⊙ = 61.1°				
	0	45	90	135	180	0	45	90	135	180	0	45	90	135	180
0	0.072	0.072	0.072	0.072	0.072	0.092	0.092	0.092	0.092	0.092	0.115	0.115	0.115	0.115	0.115
15	0.103	0.090	0.071	0.062	0.060	0.140	0.119	0.091	0.074	0.069	0.181	0.154	0.112	0.089	0.083
30	0.176	0.126	0.077	0.061	0.058	0.270	0.171	0.097	0.072	0.068	0.370	0.227	0.120	0.084	0.076
45	0.400	0.198	0.092	0.073	0.075	0.614	0.264	0.111	0.085	0.084	—	0.329	0.134	0.093	0.089
60	—	0.335	0.125	0.103	0.115	—	0.421	0.149	0.118	0.128	—	0.462	0.172	0.130	0.131
75	—	0.592	0.224	0.198	0.223	—	0.676	0.258	0.226	0.256	—	0.686	0.284	0.249	0.262

λ = 0.593; p = 0.79 ± 0.01; τ = 0.24

z°	z_⊙ = 73.9°					z_⊙ = 67.1°					z_⊙ = 59.1°				
	0	45	90	135	180	0	45	90	135	180	0	45	90	135	180
0	0.120	0.120	0.120	0.120	0.120	0.156	0.156	0.156	0.156	0.156	0.200	0.200	0.200	0.200	0.200
15	0.173	0.152	0.123	0.106	0.102	0.230	0.198	0.153	0.127	0.120	0.312	0.263	0.194	0.157	0.148
30	0.289	0.208	0.127	0.104	0.102	0.404	0.282	0.131	0.126	0.119	0.600	0.382	0.210	0.142	0.140
45	0.615	0.207	0.153	0.126	0.131	0.917	0.417	0.189	0.146	0.148	—	0.505	0.231	0.166	0.161
60	—	0.512	0.207	0.185	0.204	—	0.637	0.253	0.211	0.224	—	0.739	0.290	0.228	0.235
75	—	0.857	0.357	0.331	0.380	—	0.957	0.396	0.378	0.418	—	1.03	0.466	0.417	0.450

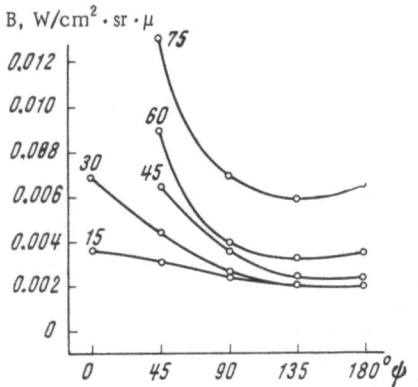

Fig. 1. Azimuthal brightness variation on almucantars (z = 15, 30, 45, 60, 75°) for $z_\odot = 61°$, $\lambda = 0.553\,\mu$, p = 0.82 (27 August 1965).

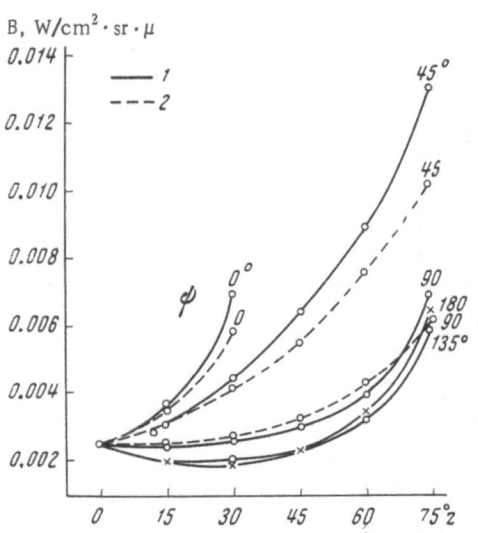

Fig. 2. Variation of observed (1) and theoretical (2) brightnesses in various vertical circles. Theory: $z_\odot = 60°$, $\tau = 0.2$, q = 0.2 (indicatrix VI). Experiment: $z_\odot = 61°$, $\tau = 0.2$, $\lambda = 0.553\,\mu$, q $\simeq 0.2$ (27 August 1965).

where B_s is the brightness at a point on the solar almucantar at a fixed angular distance from the sun ($\theta = 20°$ in our case), $m_\odot$ the atmospheric mass toward the sun, p the atmospheric transmission coefficient, and C a constant.

We used days on which the instability factor $\varkappa$, as determined by the following formulas (2) and (3), did not exceed 1.5%:

$$p \pm \Delta p = \frac{p_B + p_s}{2} \pm \left| \frac{p_s - p_B}{2} \right|, \qquad (2)$$

$$\varkappa = \frac{\Delta p}{p} = \frac{|p_s - p_B|}{p_s + p_B}. \qquad (3)$$

The brightnesses were calculated by the formula

$$B_\lambda = \frac{n_s}{n_e} \frac{A_\lambda}{\pi} \pi S_{0\lambda} p_\lambda^{m_\odot} \frac{r_0^2}{r^2}, \qquad (4)$$

where B_λ is the sky brightness in W/cm² · sr · μ; n_s and n_e the instrument readings for the observed sky point and for the screen, respectively; A_λ the spectral albedo of the gypsum screen; $\pi S_{0\lambda}$ the solar constant, which was taken from [6]; p_λ the atmospheric transmittance; $m_\odot$ the atmospheric mass toward the sun; and r_0^2/r^2 is a correction for seasonal variation of the earth-to-sun distance.

The error for brightnesses determined by formula (4) varies from observation to observation and depends upon many factors: the instrument readings for the sky and for the gypsum screen, the solar zenith distance (the greater this distance, the greater the error in determining atmospheric mass), the range of variation of atmospheric transmittance. If we find the error in B_λ by the formula:

$$\delta B_\lambda = \pm \sqrt{\delta n_s^2 + \delta n_e^2 + \delta A_\lambda^2 + \delta p_\lambda^2 m_\odot^2 + \ln^2 p_\lambda \delta m_\odot^2 m_\odot^2} \qquad (5)$$

and use the mean errors for the values in this formula, we obtain $\delta B_\lambda \simeq 5\%$. The error for $\pi S_{0\lambda}$ is ignored here.

The measurement results are presented in Table 1. Since the obtained brightnesses were entirely symmetric relative to the solar vertical circle (with the exception of the great zenith distance z = 75°), data for only half of the sky are given. The following symbols are employed in the table: λ for wavelength; p for atmospheric transmittance; τ for optical thickness, as calculated by the formula $\tau = -\ln p$; $z_\odot$ for solar zenith distance; z for the zenith distance of the observed point; and Ψ for the azimuth angle of the observed point, as read from the solar vertical circle.

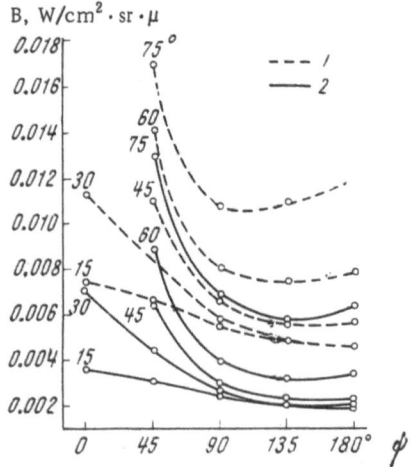

Fig. 3. Azimuthal brightness variation for several almucantars:
1) $\lambda = 0.404\,\mu$, p = 0.66, $z_\odot =$
58.7°; 2) $\lambda = 0.553\,\mu$, p = 0.82, $z_\odot =$
61° (27 August 1965).

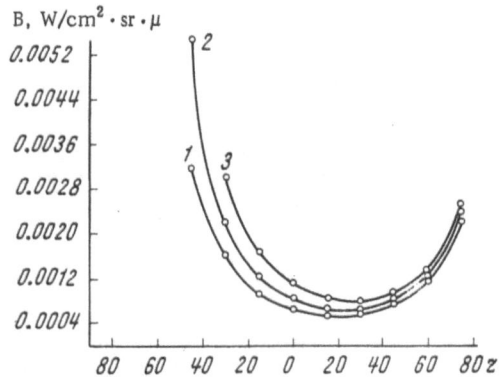

Fig. 4. Brightness variation in solar vertical for a number of sun positions at $\lambda =$
0.691 μ and p = 0.86 (27 August 1965): 1)
$z_\odot = 78°$; 2) $z_\odot = 70.4°$; 3) $z_\odot = 63.3°$.

The obtained data allow us to draw definite conclusions about regularities in the brightness distribution. The azimuthal brightness variation for different almucantars and the brightness variation for different vertical circles are shown in Figs. 1 and 2.

The position of the sun was unchanged in both the first and second cases.

The figures show that brightness decreases with an increase in azimuth angle Ψ when z = const and increases with an increase in z (toward the horizon) when Ψ = const. The minimum on the curves is due to the singularities of the scattering functions (minimum scattering at $\theta \simeq 90°$). Figure 3 shows the azimuthal brightness variation for a number of almucantars for $\lambda_1 = 5530$ Å and $\lambda_2 = 4040$ Å. If for each individual almucantar we calculate the ratio of the brightness at the point with $\Psi = 45°$ to the minimum observed brightness on the given almucantar, we see that the ratio increases with wavelength. Such ratios for almucantars z = 15, 30, 45, 60, and 75° and for two wavelengths are given below:

z	15	30	45	60	75°
$\lambda_1 = 4040$ Å	1.44	1.83	1.97	1.91	1.62
$\lambda_2 = 5530$ Å	1.63	2.32	2.78	2.78	2.24

The increase in the ratio $B_{\Psi = 45°}/B_{min}$ with wavelength may be explained by a variation in the degree of elongation of the indicatrices with wavelength.

A typical brightness distribution for the solar vertical circle at various solar zenith distances for a single wavelength is shown in Fig. 4. A brightness minimum in the antisolar vertical circle and a variation in the position of the minimum with $z_\odot$ can be seen. The rather large vertical-circle step ($\Delta z = 15°$) made it impossible to establish with high accuracy the position of the minimum-brightness point, but it is quite possible to follow the regular variation of its coordinates (see Fig. 4).

Table 2 gives the angular distances of the minimum-brightness point from the sun Δ and the zenith distances of the minimum-brightness point z for various $z_\odot$ and λ.

A. I. IVANOV

TABLE 2

16 August 1965, A. M.				27 August 1965, A. M.				14 September 1965, A. M.			
$z_\odot^\circ$	Δ°	z°	λ, Å	$z_\odot^\circ$	Δ°	z°	λ, Å	$z_\odot^\circ$	Δ°	z°	λ, Å
67.1	82.1	15	4040	72.5	85	12.5	4040	72.4	87.4	15	4470
59.6	81.6	22	4040	65.8	86	20.2	4040	65.7	82.2	16.5	4470
48.7	75.7	27	4040	58.7	81	22.3	4040	56.1	81.1	25	4470
72.5	92.5	20	6910	78	98	20	6910	76.6	92	15.4	6910
62.1	87.1	25	6910	70.4	95	24.6	6910	69.6	89	19.4	6910
54.2	84.2	30	6910	63.3	90	26.7	6910	60.0	85	25	6910

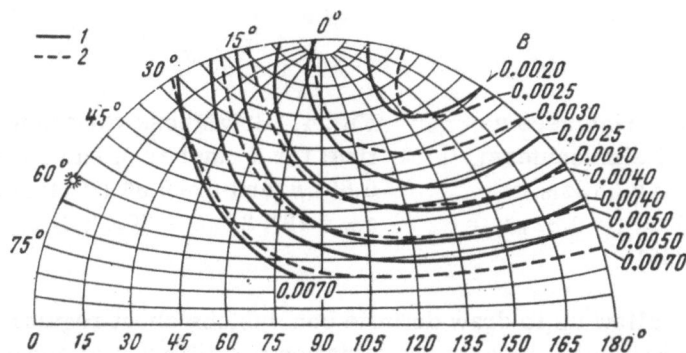

Fig. 5. Observed (1) and theoretical (2) isophots of daytime clear sky. Observation: $z_\odot = 61°$, $\lambda = 0.553\mu$, $p = 0.82$, $\tau = 0.20$, $q \simeq 0.2$ (27 August 1965). Theory: $z_\odot = 60°$, $\tau = 0.20$, $q = 0.2$, for indicatrix VI (ignoring ozone absorption).

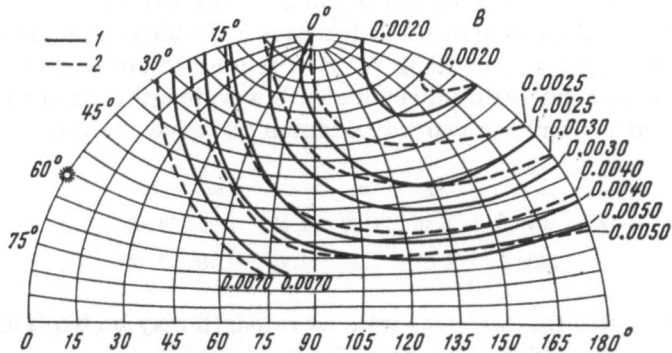

Fig. 6. Observed (1) and theoretical (2) isophots of daytime clear sky. Same parameters as in Fig. 5, but ozone absorption has been taken into account for theory.

The following may be said concerning the position of the minimum-brightness point.

1. The distance between the sun and the minimum-brightness point varies within 76-98°.

2. As the sun ascends, Δ decreases, but z (the zenith distance for the minimum-brightness point) increases, i.e., the point descends.

3. Distance Δ is a function of λ; the greater λ the greater Δ.

This latter point is at variance with the results of theoretical calculations [4] but agrees with I. N. Yaroslavtsev's data [2], which were obtained from direct brightness observations.

Some of the regularities of the brightness distribution that were mentioned above are confirmed in general outline by theoretical calculations [4, 5].

Figures 5 and 6 are presented as an example of detailed comparison of direct observations and theory. Observed and theoretical isophots for the daytime clear sky are given in Fig. 5. Inasmuch as neither the albedo q for the given locality nor the scattering indicatrix was determined experimentally, we took the most probable values of these parameters, considering the state of the underlying surface (herbage) and the spectral region in which we worked.

It can be seen from the figures that the shape of the isophots is approximately the same, but the brightnesses differ considerably in absolute value. Figure 6 shows observed and theoretical isophots for the same parameters as in Fig. 5, but the theoretical data have been corrected for ozone absorption by G. Sh. Livshits' method [3]. This correction brought the theoretical values considerably closer to the observed values, and for regions far-removed from the sun they agree with the observed values within the measurement error for most points. The increasing difference between the theoretical and observed brightnesses as we approach the sun is well-illustrated by Fig. 2. There, the theoretical and observed curves for vertical circles with $\Psi = 135$ and $180°$ merge, while for vertical circles close to the solar vertical circle ($\Psi = 0$ and $45°$), the deviations increase rapidly as z increases (toward the sun). The above examples indicate a difference between the actual and theoretical indicatrices for small scattering angles.

My sincere thanks to G. Sh. Livshits for daily assistance in this work.

LITERATURE CITED

1. E. V. Pyaskovskaya-Fesenkova, Investigation of Light Scattering in the Earth's Atmosphere [in Russian], Izd. Akad. Nauk SSSR, Moscow (1957).
2. I. N. Yaroslavtsev, The Sky's Brightness Distribution, Izv. Akad. Nauk SSSR, seriya geofiz., No. 1 (1953).
3. G. Sh. Livshits, "Light scattering in the atmosphere," Trudy Astrofiz. inst. Akad. Nauk Kazakh SSR, Vol. 6 (1965).
4. E. M. Feigel'son, M. S. Malkevich, S. Ya. Kogan, G. D. Koronatova, K. S. Glazova, and M. A. Kuznetsova, Calculation of the Brightness of Light in the Case of Anisotropic Scattering, Part 1, Consultants Bureau, New York (1960).
5. V. S. Atroshenko, E. M. Feigel'son, K. S. Glazova, and M. S. Malkevich, Calculation of the Brightness of Light in the Case of Anisotropic Scattering, Part 2, Consultants Bureau, New York (1963).
6. F. S. Jonson, "The solar constant," J. Meteorol., Vol. 2, No. 6 (1954).
7. P. N. Boiko, "An automatic-recording photoelectric spectrophotometer," Izv. Astrofiz. inst. Akad. Nauk Kazakh SSR, No. 8 (1959).

SPECTROPHOTOMETRIC STUDIES OF ATMOSPHERIC TRANSMITTANCE AND STABILITY

B. T. Tashenov

Information on the optical stability of the atmosphere is necessary for solving a wide variety of problems in atmospheric optics, particularly in the study of spectral transmittance. A number of methods for monitoring stability have been developed by V. G. Fesenkov and E. V. Pyaskovskaya-Fesenkova [1, 2]. Experimentally, the methods have mainly been tested in comparatively broad, filtered spectral regions. Little attention has been devoted to the unstable state, to the relationships and factors governing optical perturbations in the various spectral regions.

The studies of spectral transmittance and stability described here were made in the visible region (410-735 mμ) with a photoelectric spectrophotometer. The apparatus included a UM-2 monochromator on an azimuth mounting, an EPP-09 recording potentiometer, and an FEU-51 photomultiplier. The spectrophotometer (with a loop oscillograph as the recorder) has been described in detail by P. N. Boiko [3], upon whose suggestion it was built in the workshops of the Astrophysics Institute of the Academy of Sciences of the Kazakh SSR. The observations were made at the institute's observatory (Kamenskoe Plateau, 1450 m above sea level). The observatory is situated to the south of Alma-Ata and rises about 0.5 km above the city.

Our purpose was to test these methods of stability monitoring and to use them to study the spectral optical characteristics of atmospheric aerosol and their variations under quasi-stable and unstable conditions. The method of monotoring the stability of the atmosphere's optical properties, which is described in detail in [2], amounts to measuring the brightness of the solar aureole B (or of any point on the solar almucantar) and the illumination E by perpendicular solar rays at the observation point. In the presence of atmospheric stability, the diurnal variation of B/E as a function of the atmospheric mass toward the sun $m_\odot$ is expressed graphically by a straight line passing through the coordinate origin. With a gradual change in transmittance, this line no longer passes through the coordinate origin. (This is discussed at greater length in [2], and also in [4].) The method has another form: when atmospheric conditions are constant, the directional-scattering factor for the entire height of the atmosphere must remain constant:

$$\mu(\varphi) = \frac{B(\varphi)}{Em_\odot},$$ (1)

where $B(\varphi)$ is the brightness of the aureole or of any point on the solar almucantar at angular distance φ from the sun.

70

Our observations show that this method makes it possible to judge very clearly the nature of the changes that occur when atmospheric stability is disrupted gradually or suddenly. These methods are effective for the entire visible region with various kinds of underlying surfaces (snow cover and herbage).

Although $\mu(\varphi)$ allows us to judge the nature of transmittance variations, it does not permit us to assess numerically the limits of these variations. This problem can be solved by G. Sh. Livshits' method [5], which employs a formula that relates the brightness of any point on the solar almucantar to atmospheric transmittance:

$$\log \frac{B(\varphi)}{m_\odot} = m_\odot \log p + C, \tag{2}$$

where C is a constant and p is the transmission coefficient. It can be seen from formula (2) that the tangent of the slope of the line gives the value of the transmission coefficient p_s. With a stable atmosphere, the method must give a p_s that agrees within the experimental error with the transmission coefficient as determined by the Bouguer method p_B. With an unstable atmosphere, p_B and p_s will give the upper and lower limits of the transmission coefficient, and also its mean value:

$$p_m = \frac{p_B + p_s}{2} \pm \left| \frac{p_B + p_s}{2} \right|.$$

As a quantitative stability characteristic, G. Sh. Livshits [5] has proposed the so-called instability factor:

$$\varkappa = \frac{p_B - p_s}{p_B + p_s} \cdot 100\%,$$

which expresses the transmittance variation in percent and allows us to see the nature of the atmosphere's spectral stability.

Our observations have shown that the Livshits method works for the entire visible region (735–410 mμ). Table 1 gives p_B and p_s for a number of wavelengths and for various days. On especially stable days, the agreement between p_B and p_s is good (the error, evidently experimental, is not over 1.0%). Even on the same day, the instability factor can vary according to wavelength, which indicates that the aerosol spectrum is variable. Table 2 gives instability factors determined from p_B and p_s observations on the aureole ($\varphi = 2°$) and at $\varphi = 57°$. The aureole observations gave considerably higher instability factors than did the observations at $\varphi = 57°$. This is entirely natural, since the course of the line $\log(B/m_\odot) = f(m_\odot)$ for the aureole is chiefly determined by the coarse aerosol fraction, which causes the corresponding limit of transmittance variation to be overestimated.

In the visible region, aerosol is almost entirely responsible for variation in the optical properties of the atmosphere (to a certain extent, ozone is also responsible, but its contribution is extremely small for short time intervals).

Simultaneous investigation of the aerosol scattering factors μ_a for $\varphi = 2°$, calculated for the entire height of the atmosphere, and their spectral variation, enables us to judge the variations in atmospheric aerosol. We calculated $\mu_a(2°)$ by E. V. Pyaskovskaya-Fesenkova's method [2]:

$$\mu_a = \mu - \left(\mu_{57°} - \frac{\tau_B - \tau_{oz}}{4\pi} \right) - \mu_R,$$

B. T. TASHENOV

TABLE 1

λ, mμ	23 December 1964, A. M.					4 January 1965, A. M.					4 January 1965, P. M.				
	p_B	p_s	Δp	p_m	x, %	p_B	p_s	Δp	p_m	x, %	p_B	p_s	Δp	p_m	x, %
735	0.906	0.875	0.016	0.890	1.8	0.906	0.897	0.004	0.902	0.4	0.902	0.794	0.050	0.848	5.9
691	0.904	0.881	0.014	0.892	1.6	0.891	0.873	0.009	0.882	1.0	0.902	0.794	0.050	0.848	5.9
650	0.885	0.849	0.018	0.867	2.1	0.875	0.873	0.001	0.874	0.0	0.885	0.809	0.038	0.847	4.5
593	0.832	0.782	0.025	0.807	3.1	0.847	0.828	0.010	0.840	1.2	0.855	0.755	0.050	0.805	6.3
553	0.832	0.794	0.019	0.813	2.3	0.778	0.828	0.025	0.800	3.1	0.832	0.745	0.044	0.788	5.6
520	0.832	0.805	0.014	0.818	1.7	—	—	—	—	—	—	—	—	—	—
508	—	—	—	—	—	0.738	0.817	0.040	0.780	5.1	0.818	0.745	0.036	0.782	4.6
495	0.826	0.789	0.018	0.808	2.2	—	—	—	—	—	—	—	—	—	—
470	0.776	0.785	0.004	0.780	0.5	0.728	0.783	0.035	0.756	7.3	0.783	0.741	0.021	0.762	2.8
447	0.773	0.736	0.018	0.754	2.4	0.708	0.762	0.054	0.735	7.4	0.757	0.708	0.024	0.732	3.3
423	0.736	0.692	0.022	0.714	3.1	0.736	0.723	0.006	0.730	0.8	0.736	0.698	0.019	0.717	2.6

λ, mμ	5 January 1965, A. M.					5 January 1965, P. M.					6 January 1965, A. M.				
	p_B	p_s	Δp	p_m	x, %	p_B	p_s	Δp	p_m	x, %	p_B	p_s	Δp	p_m	x, %
735	0.946	0.728	0.109	0.84	13.1	0.931	0.828	0.052	0.880	5.9	0.927	0.944	0.008	0.936	0.8
690	0.944	0.724	0.110	0.83	13.2	0.923	0.832	0.046	0.878	5.2	0.904	0.879	0.012	0.892	1.4
650	0.946	0.698	0.124	0.82	15.2	0.904	0.802	0.051	0.853	6.0	0.897	0.879	0.009	0.888	1.0
593	0.929	0.689	0.120	0.81	14.8	0.887	0.757	0.065	0.822	7.9	0.845	0.867	0.011	0.856	1.3
553	0.849	0.723	0.063	0.79	8.0	0.841	0.746	0.048	0.794	6.0	0.851	0.817	0.017	0.834	2.0
508	0.764	0.723	0.020	0.74	2.8	0.838	0.773	0.032	0.806	4.0	0.785	0.817	0.016	0.801	2.0
470	0.753	0.759	0.003	0.76	0.0	0.811	0.705	0.053	0.758	7.0	0.785	0.760	0.012	0.772	1.6
447	0.708	0.686	0.011	0.70	1.6	0.783	0.705	0.039	0.744	5.2	0.769	0.750	0.010	0.760	1.3
423	0.610	0.594	0.008	0.60	1.3	0.750	0.643	0.054	0.696	7.8	0.748	0.716	0.016	0.732	2.2
410	—	—	—	—	—	—	—	—	—	—	0.738	0.736	0.001	0.737	0.2

Table 1 (cont.)

6 January 1965, P. M.

λ, mμ	p_B	p_s	Δp	p_m	x, %
735	0.908	0.904	0.002	0.906	0.2
691	0.883	0.891	0.004	0.897	0.4
650	0.853	0.891	0.019	0.872	2.2
593	0.794	0.840	0.023	0.818	2.8
553	0.802	0.826	0.012	0.814	1.5
508	0.807	0.826	0.010	0.816	1.2
470	0.759	0.805	0.023	0.782	3.0
447	0.769	0.757	0.006	0.763	0.9
423	0.713	0.711	0.001	0.712	0.1
410	—	—	—	—	—

7 January 1965, A. M.

λ, mμ	p_B	p_s	Δp	p_m	x, %
735	0.933	0.847	0.043	0.89	4.9
691	0.929	0.861	0.034	0.90	3.8
650	0.883	0.815	0.034	0.85	4.0
593	0.859	0.789	0.022	0.84	2.6
553	0.836	0.789	0.023	0.81	3.5
508	0.798	0.794	0.002	0.796	0.2
470	0.752	0.736	0.008	0.744	1.1
447	0.740	0.728	0.006	0.734	0.8
423	0.713	0.708	0.032	0.710	0.3
410	0.631	0.676	0.022	0.650	3.4

4 February 1965, A. M.

λ, mμ	p_B	p_s	Δp	p_m	x, %
735	0.970	0.817	0.076	0.90	8.4
691	0.916	0.834	0.041	0.88	4.7
650	0.929	0.799	0.065	0.86	7.6
593	0.904	0.724	0.090	0.81	11.1
553	0.361	0.726	0.068	0.79	8.6
508	0.867	0.703	0.082	0.78	10.5
470	0.820	0.705	0.058	0.76	7.6
447	0.794	0.597	0.098	0.70	14.0
423	0.813	0.575	0.119	0.69	17.2
410	—	—	—	—	—

6 May 1965, A. M.

λ, mμ	p_B	p_s	Δp	p_m	x, %
735	0.772	0.802	0.015	0.785	1.9
691	0.791	0.804	0.006	0.798	0.8
650	0.791	0.794	0.002	0.792	0.3
553	0.721	0.741	0.010	0.731	1.4
520	0.700	0.724	0.012	0.712	1.7
495	0.708	0.724	0.008	0.716	1.1
470	0.676	0.687	0.006	0.682	0.9
447	0.661	0.668	0.004	0.664	0.6
423	0.631	0.621	0.005	0.626	0.8
410	0.631	0.600	0.016	0.616	2.6

12 May 1965, A. M.

λ, mμ	p_B	p_s	Δp	p_m	x, %
735	0.813	0.851	0.019	0.832	2.3
691	0.851	0.841	0.005	0.846	0.6
650	0.832	0.822	0.005	0.827	0.6
553	0.794	0.759	0.018	0.776	2.3
520	0.776	0.780	0.002	0.778	0.3
495	0.741	0.759	0.009	0.750	1.2
470	0.708	0.741	0.016	0.724	2.2
447	0.682	0.716	0.017	0.699	2.5
423	0.661	0.631	0.045	0.646	2.3
410	0.646	0.617	0.014	0.632	2.2

15 May 1965, A. M.

λ, mμ	p_B	p_s	Δp	p_m	x, %
735	0.759	0.741	0.009	0.750	1.2
691	0.759	0.741	0.009	0.758	1.2
650	0.711	0.733	0.006	0.746	0.5
553	0.708	0.692	0.008	0.700	1.1
520	0.692	0.676	0.008	0.684	1.2
495	0.692	0.676	0.008	0.684	1.2
470	0.661	0.646	0.008	0.653	1.2
447	0.631	0.631	0.000	0.631	0.0
423	0.603	0.631	0.014	0.617	2.3
410	0.592	0.550	0.021	0.571	3.5

B. T. TASHENOV

Table 1 (cont.)

λ,mμ	5 June 1965, A. M.					7 June 1965, A. M.					17 September 1965, A. M.				
	p_B	p_S	p_m	Δp	x, %	p_B	p_S	p_m	Δp	x, %	p_B	p_S	p_m	Δp	x, %
735	0.861	0.877	0.869	0.008	0.9	0.918	0.857	0.888	0.030	3.3	0.876	0.940	0.908	0.032	3.5
691	0.851	0.851	0.851	0.000	0.0	0.920	0.879	0.900	0.020	2.2	0.886	0.962	0.924	0.038	4.1
650	0.813	0.794	0.804	0.010	1.2	0.889	0.867	0.878	0.011	1.2	0.864	0.948	0.906	0.042	4.6
593	—	—	—	—	—	—	—	—	—	—	0.823	0.869	0.846	0.023	2.7
553	0.794	0.794	0.794	0.000	0.0	0.840	0.805	0.822	0.018	2.2	0.821	0.880	0.850	0.030	3.5
520	0.776	0.807	0.792	0.016	2.0	0.859	0.794	0.826	0.032	4.0	0.810	0.853	0.832	0.022	2.7
495	0.758	0.776	0.767	0.009	1.2	0.836	0.785	0.810	0.026	3.2	0.783	0.834	0.808	0.026	3.2
470	0.724	0.750	0.737	0.013	1.8	0.849	0.785	0.817	0.032	4.0	0.762	0.810	0.786	0.024	3.0
447	0.724	0.769	0.746	0.022	2.9	0.843	0.774	0.808	0.034	4.2	0.752	0.796	0.774	0.022	2.9
423	0.684	0.764	0.724	0.040	5.6	0.769	0.714	0.742	0.028	3.8	0.738	0.744	0.741	0.003	0.4
410	0.676	0.813	0.744	0.068	9.1	0.743	0.644	0.694	0.050	7.3	0.716	0.681	0.698	0.018	2.6

λ,mм	18 September 1965, A. M.					23 September 1965, A. M.					4 October 1965, A. M.				
	p_B	p_S	p_m	Δp	x, %	p_B	p_S	p_m	Δp	x, %	p_B	p_S	p_m	Δp	x, %
735	0.902	1.016	0.959	0.057	5.9	0.805	0.891	0.848	0.043	5.1	0.923	0.912	0.918	0.006	0.7
691	0.891	1.016	0.954	0.062	6.5	0.807	0.914	0.860	0.054	6.3	0.897	0.883	0.890	0.007	0.8
650	0.885	1.008	0.956	0.072	7.5	0.817	0.891	0.854	0.037	4.4	0.904	—	—	—	—
593	0.830	0.962	0.896	0.066	7.3	0.794	0.871	0.834	0.038	4.6	0.863	0.851	0.857	0.006	0.7
553	0.830	0.931	0.880	0.050	5.7	0.773	0.867	0.820	0.047	5.7	0.863	0.851	0.857	0.006	0.7
520	0.828	0.935	0.882	0.054	6.1	0.773	0.820	0.796	0.024	3.0	0.849	0.826	0.838	0.012	1.4
495	0.794	0.910	0.852	0.058	6.8	0.750	0.818	0.784	0.034	4.3	0.824	0.817	0.820	0.004	0.5
470	0.778	0.887	0.832	0.054	6.5	0.724	0.800	0.762	0.038	5.0	0.807	0.794	0.800	0.006	0.8
447	0.755	0.871	0.813	0.058	7.1	0.692	0.731	0.712	0.020	2.8	0.759	0.755	0.757	0.002	0.3
423	0.723	0.822	0.772	0.050	6.5	0.674	0.728	0.701	0.027	3.9	0.755	0.719	0.737	0.018	2.4
410	0.702	0.824	0.768	0.061	8.0	0.647	0.671	0.644	0.027	4.2	0.762	0.724	0.743	0.019	2.5

Table 1 (cont.)

λ, mμ	14 October 1965, A. M.					16 October 1965, A. M.				
	p_B	p_S	p_m	Δp	x, %	p_B	p_S	p_m	Δp	x, %
735	0.920	0.881	0.900	0.020	2.2	0.847	0.946	0.896	0.050	5.6
691	0.912	0.885	0.896	0.014	1.6	0.851	0.942	0.898	0.046	5.1
650	0.902	0.871	0.886	0.016	1.8	0.861	0.927	0.894	0.033	3.7
617	0.879	0.851	0.865	0.014	1.6	0.832	0.910	0.871	0.039	4.7
593	0.863	0.843	0.853	0.010	1.2	0.817	0.895	0.856	0.039	4.6
553	0.867	0.832	0.850	0.018	2.1	0.809	0.871	0.840	0.031	3.7
520	0.843	0.826	0.834	0.008	1.0	0.791	0.867	0.829	0.038	4.6
495	0.834	0.813	0.824	0.010	1.2	0.782	0.849	0.816	0.034	4.1
470	0.813	0.792	0.802	0.010	1.2	0.764	0.822	0.793	0.029	3.6
447	0.771	0.787	0.779	0.008	1.3	0.741	0.822	0.782	0.040	5.1
423	0.748	0.745	0.746	0.002	0.3	0.689	0.783	0.736	0.047	6.4
410	0.724	0.678	0.702	0.023	3.3	0.674	0.748	0.711	0.037	5.2

TABLE 2

λ, mμ	14 October 1965, A. M.							16 October 1965, A. M.						
	p_B	p_S (2°)	p_m	x_1, %	p_S (57°)	p_m	x_2, %	p_B	p_S (2°)	p_m	x_1, %	p_S (57°)	p_m	x_2, %
730	0.920	0.881	0.900	2.2	0.914	0.917	0.3	0.847	0.946	0.896	5.6	0.885	0.866	2.2
691	0.912	0.885	0.896	1.6	0.904	0.908	0.4	0.851	0.942	0.898	5.1	0.867	0.859	0.9
650	0.902	0.871	0.886	1.8	0.891	0.896	0.7	0.861	0.927	0.894	3.7	0.849	0.855	0.7
593	0.863	0.843	0.853	1.2	0.861	0.862	0.1	0.817	0.895	0.856	4.6	0.839	0.828	1.3
553	0.867	0.832	0.850	2.1	0.840	0.854	1.7	0.809	0.871	0.840	3.7	0.811	0.810	0.1
520	0.843	0.826	0.834	1.0	0.828	0.842	0.2	0.791	0.867	0.829	4.6	0.798	0.794	0.5
495	0.834	0.813	0.824	1.2	0.794	0.831	0.4	0.782	0.849	0.816	4.1	0.794	0.788	0.8
470	0.813	0.792	0.802	1.2	0.776	0.804	1.2	0.764	0.822	0.793	3.6	0.771	0.768	0.5
447	0.771	0.787	0.779	1.3	0.736	0.774	0.3	0.741	0.822	0.782	5.1	0.750	0.746	0.5
423	0.748	0.745	0.746	0.3	0.713	0.742	0.8	0.689	0.783	0.736	6.4	0.721	0.705	2.3
410	0.724	0.678	0.702	3.3		0.718	0.8	0.674	0.748	0.711	5.2	0.708	0.691	2.4

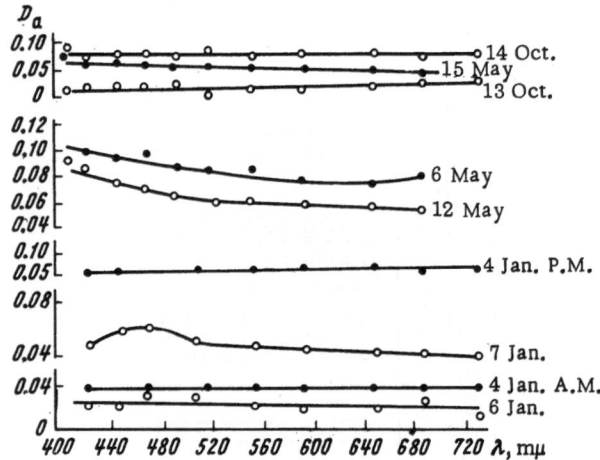

Fig. 1. Spectral optical aerosol density.

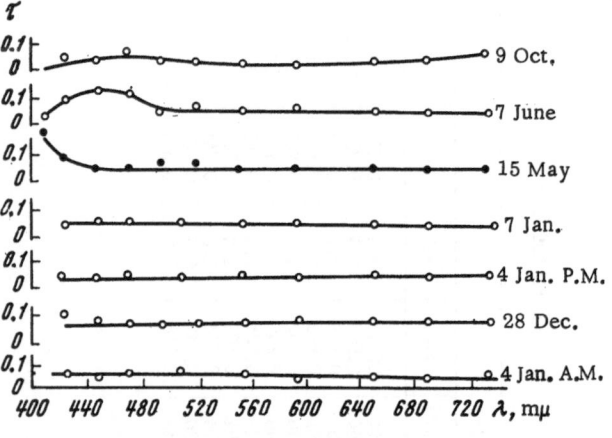

Fig. 2. Spectral optical thickness of haze.

where $\mu_{57°}$ is the total scattering factor at $\varphi = 57°$, μ_R the Rayleigh scattering factor for $\varphi = 2°$, τ_B the total optical thickness of the atmosphere, and τ_{oz} the ozone optical thickness.

First of all, it should be pointed out that three groups of stability states manifest themselves.

1) Ideal stability, when the instability factor $\varkappa \leq 1\%$ (in this case, $\varkappa$ is determined by the experimental errors in p_B and p_s).

2) Quasi-stability, when transmittance varies gradually in either direction (Bouguer lines of the relative aureole can be plotted).

3) Instability, when the optical properties of the atmosphere vary irregularly (the relative-aureole line and $\log (B/m_\odot) = f(m_\odot)$ cannot be plotted).

The first case is encountered relatively rarely; the second case is observed most often. The gradual change in the atmosphere's optical properties in our observation area is closely related to mountain–valley air circulation.

Case three occurs most often around noon, when the haze comes in from the steppe and the city, causing more or less abrupt changes in atmospheric transmittance. In most cases, the change is in the same direction for all wavelengths. Sometimes, a worsening of transmittance in some spectral regions is accompanied by an improvement or a lack of change in others (this can be seen from the instability factors in Table 1). This occurs only when there is a comparatively slow, progressive change in transmittance, when the circulating air evidently carries in particles of one size and removes particles of another size. The aerosol-particle spectrum also changes during an unstable period.

Let us consider the aerosol characteristics during quasi-stable and unstable states. The spectral variation of the aerosol scattering functions in the aureole region during quasi-stability is represented by smooth curves that rise as wavelength decreases (according to data from 12 observation days). But their curvature differs even from one day to the next, which indicates qualitative variability of the atmosphere's aerosol component. Data on the total aerosol radiation attenuation also indicate that the aerosol composition is variable. As is evident from Fig. 1, the spectral optical aerosol density D_a during a quasi-stable period: 1) can be steady; 2) can increase as wavelength decreases; and 3) can decrease as wavelength decreases.

Of the 16 series of observations, the first situation occurred in four, the second in eight, and the third in two series. The aerosol extinction curves for 7 January 1965 and for 4 October 1965, which have maxima in the areas of 470 mμ and 450 mμ, respectively, are of particular interest. It should be noted that no maximum is observed in $\mu(\lambda)$ for $\varphi = 2°$ or for

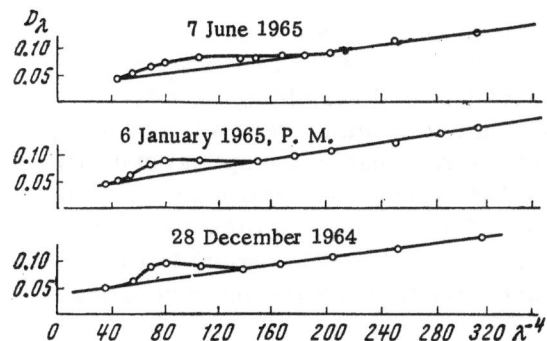

Fig. 3. Isolation of absorption band for ozone determination.

$\varphi = 57°$ in the vicinity of 450–470 mμ, which shows that the aerosol–extinction peak near 470 mμ is due to absorption as well as scattering.

The spectral variation of aerosol optical density is often represented as $D_a = k/\lambda^n$, where k and n are constants. The n values determined from 12 series of observations were in the range of 0.00–1.40, with a mean value $n_m = 0.34$. Negative values of n were found in two cases, i.e., aerosol extinction increases with an increase in wavelength. It must be mentioned here that the observations on the afternoon of 4 January (n = −0.22) were made after the haze had arrived (on the morning of 4 January, n = 0.0), and there had been a dust storm for two days prior to 13 October (n = −1.42). These factors evidently produced coarsely dispersed aerosol (and the respective n values) on these days. Reduced n was also noted during the second half of 6 May 1965, after the haze had set in.

There are also days for which it is difficult to attribute a particular n value to aerosol extinction. On 6 May 1965, for example, n = 0.0 for 735–593 mμ, but n = 0.83 for 593–410 mμ. On 12 May 1965, n = 0.62 for 735–520 mμ and n = 1.40 for 520–410 mμ. There were days (7 January and 4 October), too, when the aerosol optical density had a peak in the area of 450–470 mμ.

Similar data were obtained for aerosol haze. Figure 2 shows the spectral variation of the optical thickness of the haze and curves of $\mu_h(2°) = \mu_h(\lambda)$. These data ($\tau_h$ and μ_h) were found by subtracting the prehaze values τ_a and μ_a from the data obtained after the arrival of the haze.

The nature of the spectral variation of $\mu_h(2°)$ varied even within a single day, which indicates the diversity of the particle spectra of the hazes that arrived or departed at various times. Three forms of the relation $\mu_h(2°) = \mu_h(\lambda)$ may be distinguished.

1. Increase in $\mu_h(2°)$ with decrease in wavelength (19 cases out of 25).

2. No change (three cases out of 25).

3. Decrease in $\mu_h(2°)$ with decrease in wavelength (three cases out of 25).

It was interesting to compare the shape of the curves for $\mu_h(2°) = \mu_h(\lambda)$ and $\mu_a(2°) = \mu_a(\lambda)$ for the same day. Our analysis showed that for four out of six days, the variation of $\mu_a(2°)$ with wavelength was more monotonic than that of $\mu_h(2°) = \mu_h(\lambda)$. On the other two days, the relation $\mu_h(2°) = \mu_h(\lambda)$ was the more monotonic. This indicates that the relative number of fine haze particles, as compared with spectrum of the aerosol distributed in the atmosphere during a stable period, can vary in either direction. This is also demonstrated by a comparison of the spectral optical thicknesses of haze and aerosol during a stable period.

In the majority of cases, the optical thickness of the haze increased monotonically as wavelength decreased or was indifferent to it (neutral). The data on haze optical thickness for 7 June and 9 October 1965, when a rather clearly expressed maximum in the area of 450 and 470 mμ (evidently due, as has already been noted, to absorption) was observed, are also quite interesting. The shape of the curve for $\tau_h = \tau_h(\lambda)$ for 15 May 1965, which shows an appreciable increase in haze optical thickness at 410–440 mμ, is typical.

Monitoring the stability of the optical properties of the atmosphere makes it possible to assess the effect of variation in atmospheric conditions on the errors of the final results. We evaluated the accuracy in measuring spectral transmittance and in determining the effective ozone thickness from the absorption bands. The accuracy of determining transmittance by the Bouguer method is primarily governed not by experimental errors but by the optical stability of the atmosphere. Data on mean transmittances p_m and their respective errors are given in Table 1. It can be seen that even for one day the accuracy of p_m varies according to wavelength λ (because the degree of instability varies according to the spectral region) and is within 5%. But there are days when the deviations from p_m reach 10-15% (the atmosphere's optical properties varied rapidly but smoothly on these days).

Let us consider the accuracy of determining ozone thickness from the absorption bands, because here the errors are connected with the accuracy of determining transmittance over the entire visible region. If we plot the line $D_\lambda = D(\lambda^{-4})$ (D is the total atmosphere optical density), it will have a hump resulting from radiation attenuation in the absorption band (Fig. 3). The error in the equivalent ozone thickness x depends upon the error in $D = -\log p$, i.e., of the error in p. Thus, even on a stable day ($\varkappa = 0.5\%$) the accuracy of x is no better than 10%. On less stable days, the accuracy is correspondingly lower. It can be seen from the data below that ozone thickness varies within rather wide limits (4.9-1.6 mm). But this is typical during a change of season; it usually varies little within one month.

Date	Ozone	Date	Ozone
28 December 1964, A. M.	4.9 ± 0.7	15 May 1965, A. M.	1.8 ± 0.4
4 January 1965, A. M.	1.6 ± 0.4	5 June 1965, A. M.	4.9 ± 0.7
6 January 1965, A. M.	4.1 ± 0.4	7 June 1965, A. M.	3.3 ± 0.8
6 May 1965, A. M.	1.6 ± 0.4	4 October 1965, A. M.	2.3 ± 0.3
12 May 1965, A. M.	1.6 ± 0.4	14 October 1965, A. M.	1.6 ± 0.3

Summary

1. The methods that we have examined for monitoring optical stability are effective for the entire visible region and various underlying surfaces. They enable us to study the regularities of the unstable state and to assess the effect of atmospheric perturbations on the accuracy of parameter determinations such as transmittance and effective ozone thickness.

2. Three stability states can be distinguished: a) stable (relatively rare); quasi-stable, in which transmittance changes gradually (most frequent); and c) unstable, in which transmittance variation is nonmonotonic (occurs mainly at the moment of arrival of dense haze).

Changes in the atmosphere's optical properties in the visible region are caused by quantitative as well as qualitative aerosol fluctuations.

3. During a quasi-stable period, aerosol optical density D_a varies with wavelength in three ways: a) it increases as λ decreases (observed in eight series out of 16); b) it is indifferent to wavelength (in four series out of 16); and c) it decreases as λ decreases (in two series out of 16). A peak in the area of 450-470 mμ was observed on two days.

Frequently, the aerosol optical density can be represented as $D_a = k/\lambda^n$. For our observations, n varied within 0.0-1.40 with a mean value $n_m = 0.34$. In two cases, n took a negative value (n $= -0.22$ and -1.42), and in two other cases, it varied according to the spectral region.

In a quasi-stable period, the aerosol directional-scattering factor for $\varphi = 2°$ increases as λ decreases, and the curvature of the $\mu_a(2°) = \mu_a(\lambda)$ lines varies from day to day (12 days of observations).

4. The particle spectra of arriving and departing haze, even for the same day, are quite diverse, but well-defined groups can be distinguished for optical thickness as well as for the directional-scattering factors $\mu_h(2°)$.

The haze optical thickness τ_h increases monotonically with a decrease in wavelength λ (in three out of eight observation series) or is neutral (in three series out of eight). A peak in $\tau_h = \tau_h(\lambda)$ in the area of 450–470 mμ was observed on two days.

The directional-scattering factor $\mu_h(2°)$ is found to: a) increase as λ decreases (19 cases out of 25); b) be neutral (three cases out of 25); and c) decrease as λ decreases (three cases out of 25).

5. Comparison of $\mu_h(2°) = \mu_h(\lambda)$ with $\mu_a(2°) = \mu_a(\lambda)$ and $\tau_h = \tau_h(\lambda)$ with $\tau_a = \tau_a(\lambda)$ shows that the haze particle spectrum can have a greater or smaller coarse fraction than the aerosol spectrum during a stable period.

I thank G. Sh. Livshits for suggesting the topic and for guidance.

LITERATURE CITED

1. V. G. Fesenkov, Astron. zh., Vol. 10, No. 3 (1933).
2. E. V. Pyaskovskaya-Fesenkova, Investigation of Light Scattering in the Earth's Atmosphere [in Russian], Izd. Akad. Nauk SSSR, Moscow (1957).
3. P. N. Boĭko, Izv. Astrofiz. inst. Akad. Nauk Kazakh SSR, No. 8 (1950).
4. G. Sh. Lifshits, Izv. Astrofiz. inst. Akad. Nauk Kazakh SSR, Vol. 5, No. 7 (1957).
5. G. Sh. Livshits, Trudy Astrofiz. inst. Akad. Nauk Kazakh SSR, No. 6 (1965).

LIGHT POLARIZATION AS AN INDEPENDENT MEASURE OF ATMOSPHERIC PURITY AND TRANSMITTANCE

D. G. Stamov

The purity of the atmosphere is characterized by the quantity and size of the particles of dust and moisture (in liquid and solid phases) suspended in it. That the purity, or turbidity, of the atmosphere can be assessed from the polarizing properties of its gaseous and aerosol components is a point that does not meet with any fundamental objections. But the practical solution of this important problem involves great difficulties. And this explains the conflicting opinions that have been expressed in recent years by many researchers [1, 2].

Any optical determination of atmospheric turbidity is liable to be based on relationships that exist between turbidity and absorption and scattering processes. It is to characterize these phenomena that the transmission coefficient and degree of polarization were introduced. Both values are entirely limited by absorption and scattering of direct and reflected sunlight. Both processes are controlled by the more or less constant gaseous and highly variable aerosol components of the atmosphere.

It has been firmly established, both experimentally and theoretically, that aerosol, which is ever present in the air at all altitudes, has a much greater effect on light absorption and scattering than do air-molecule fluctuations. Without taking into account the manifold properties (including optical) of aerosols, it is impossible to analyze the intricacies of the phenomena that occur in the real atmosphere. The notion of an ideal (dry and dustfree) atmosphere is an abstraction from which researchers are attempting to free themselves.

Aerosol is ununiformly distributed in the atmosphere, is constantly being shifted by vertical and horizontal air currents, and is always undergoing qualitative changes, especially as a result of condensation processes. Liquid-phase water is among the most variable components of the atmosphere. Aerosols are almost entirely responsible for the variability of atmospheric transmittance and polarization. This important conclusion follows from many atmospheric optics studies made over the past 15–20 years, mainly involving development of the theory of radiation transfer in the atmosphere [1].

Since quantitative estimates of transmittance and turbidity have been introduced into practice on the basis of actinometric and photometric considerations, in polarimetry it was necessary to show, first of all, that these estimates were linearly related to polarization. Attempts have been made to establish the existence of such a relationship by statistical processing of simultaneous actinometric (or photometric) and polarimetric observations. As might be expected, in the process it was found that the relationship between polarization and actinometric (or photometric) evaluations of transmittance and turbidity was not functional (one-to-one) but correlative (stochastic). A single polarization value corresponded not to one but to a wide

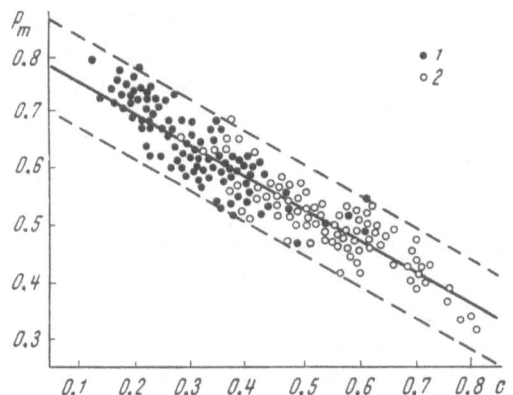

Fig. 1. Correlation between maximum polarization and total turbidity factor after Kastrov: 1) maximum polarization from almucantar measurements by Stamov in autumn 1958; 2) maximum polarization from measurements on solar vertical circle by Serpitskaya in summer 1955.

variety of transmission-coefficient (or turbidity-factor) values, and vice versa [3]. There is nothing unexpected in this fact (Fig. 1): it is a quite natural consequence when the values being compared are not exactly comparable. In the first place, they are characteristics of different phenomena, and secondly, besides turbidity, they are functions of a number of other factors, many of which they do not have in common [4]. The correlation coefficient for polarization and turbidity (or transmittance) has not been and will not be unity — it is on the order of 0.6-0.8. And inasmuch as a functional relationship does not exist, substituting a polarimetric characteristic for an actinometric (or photometric) characteristic is out of the question. But one optical characteristic can be determined from another, and their probable values and deviations in one direction or another can be indicated. Figure 1 shows that when turbidity is determined from polarization this deviation is approximately ±0.15, and when polarization is determined from turbidity it is about ±0.10.

The Kastrov total turbidity factor c, which is expressed in terms of the solar constant S_0 and the solar-radiation intensity S on the earth's surface for atmospheric mass m,

$$c = \frac{S_0 - S}{mS},$$

shows the attenuating effect of the atmosphere on solar radiation by pure absorption as well as by scattering. It is essential to note that c allows us to assess turbidity only toward the sun; the turbidity in other directions remains undetermined.

Polarization, on the other hand, is a very sensitive indicator of the scattering of sunlight by air-molecule fluctuations and aerosols in any direction [6, 10]. The following factors, which govern the polarization of light scattered back toward the observer, are discussed in detail in [7]. To the light that is singly scattered by air-molecule fluctuations (factor I) is added the light that is singly scattered in the same direction by aerosols (factor II), followed by the light that is multiply scattered by molecules and aerosols (factor III), and, finally, the light reflected from the earth's surface that is scattered by them (factor IV). Calculation of scattering of any order must also take into account the elongation of the scattering indicatrix (factor V), extinction (factor VI), the optical anisotropy of molecular scattering (factor VII), and the aerosol size distribution (factor VIII).

Theoretical allowance for all of these factors is very complicated. A theory of light scattering in the real atmosphere has yet to be created. There is a theory of light scattering only for an absolutely pure (dry and dustfree) atmosphere.

Research indicates that factors III and IV reduce polarization while factors V and VI increase it. This compensation (although it is not total) allows factors III through VI to be ignored in the first approximation, so that factors I and II may be considered controlling (or, in any case, dominant). On this basis, researchers resort to rough estimates, considering only factors I and II and ignoring the rest.

Proceeding from the fact that single scattering was caused by air molecules as well as large particles suspended in the atmosphere and assuming that molecular scattering polarized

light almost completely while large particles scattered light almost indifferently, in 1925 Milch
[5], and then independently Tikhanovskii (1927) and myself (1944), derived the following semiempirical approximation formula relating daytime clear-sky polarization to atmospheric turbidity:

$$p = \frac{\sin^2 \theta}{2 - \sin^2 \theta + F},$$ (1)

where θ is the scattering angle and F is a parameter whose numerical value depends upon the
nature and degree of turbidity of the well-mixed scattering medium – the atmosphere [7]. For
a medium consisting of isotropic molecules, we let F = 0 and obtain the Rayleigh formula. But
if we allow for anisotropy a, we must let $F = 2a/(1 - a) \simeq 0.1$, and we obtain the Cabannes –
Tikhanovskii formula. When the scattering medium contains large particles as well as molecules, the parameter F takes values exceeding 0.1, according to the size, properties, and number of these particles. The more dust particles and condensed moisture in the atmosphere,
the greater the numerical value of F and the less the polarization, as can be seen from direct
observations.

The main point of the above semiempirical approximation formula is that it offers a
graphic demonstration of all polarization variations at various points in the daytime clear sky
as functions of the turbidity of a well-mixed atmosphere. In any locality, for any hour, day, or
season, and at any scattering angle, polarization increases as turbidity decreases because of
meteorological conditions and the synoptic situation [8, 9, 11].

The depolarizing effect of turbidity considerably exceeds the effect of all other depolarizing factors. Turbidity is the common factor that affects both the intensity and polarization of
sunlight in all layers of the atmosphere and throughout its entire thickness. The parameter F,
therefore, can and must be used as the new, reliable measure of turbidity, as determined from
almucantar measurements of polarization.

It must be pointed out, however, that a thorough experimental check has shown me
that formula (1) gives polarization values that are in very good agreement with almucantar observations only during sunrise, sunset, and bright twilight, i.e., when the sun is
below 6°. It was found that when the sun's elevation is $0° \le h_\odot \le 9°$, the parameter F does not
depend solely upon turbidity, and this can be taken into account by the empirical formula

$$F = W (1 + \cot^2 \theta \sin^2 10 h_\odot).$$ (2)

For $h_\odot \ge 9°$, we find

$$F = \frac{W}{\sin^2 \theta}.$$ (3)

The coefficient W in formulas (2) and (3) stands for a new parameter that depends solely upon
turbidity and not upon the scattering angle or the sun's elevation. When $\theta = 90°$, we have

$$F = W = \frac{1 - p_m}{p_m}.$$ (4)

Thus, to determine the turbidity polarization parameter, it is sufficient to measure the
maximum polarization p_m on the solar vertical circle for small elevations and on the solar
almucantar for large elevations, bearing in mind that maximum polarization is somewhat dependent upon the inclination of the line of sight relative to the horizon.

The relationship between W and c can be expressed by the very approximate formula

$$W = \frac{0.2 + c}{1.2 - c},$$ (5)

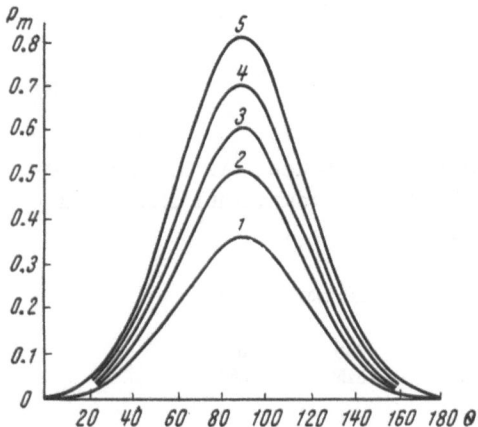

Fig. 2. Degree of polarization as a function of scattering angle and turbidity, obtained by author in 1958 from almucantar measurements of polarization and simultaneous actinometric measurements of turbidity: 1) 17 May, c = 0.71; 2) 12 September, c = 0.45; 3) 27 September, c = 0.35; 4) 19 September, c = 0.20; 5) 20 September, c = 0.14.

which was derived by statistical processing of simultaneous determinations of p_m and c [10].

We shall not touch upon all of the details involved with the use of these formulas or demonstration of their practical value. This can be found in [10-12].

Many of my studies lead me to conclude that the atmosphere's aerosol component has a double polarizing effect, as Sekera has already correctly observed [1]. On the one hand, its action (chiefly that of its fine fraction, of course) is patterned after that of the molecular component, causing a Rayleigh-like symmetric angular distribution of polarization (Fig. 2). On the other hand, its coarse fraction causes a number of deviations from the classical Rayleigh theory. It reduces polarization in almost the same respect for all scattering angles, shifts the polarization maximum several degrees from the sun, displaces the neutral points from the sun and antihelion, and elongates the forward-scattering indicatrix.

It was shown above how the author used certain of the properties of the aerosol component to introduce a polarization characteristic for atmospheric turbidity. S. I. Sivkov used the same properties to introduce another characteristic, which he called "turbidity depolarization" [13]. First of all, he focused his attention on the fact that the difference between theoretical (calculated for an absolutely pure atmosphere) and observed (measured in the real atmosphere) not only represented a deviation of theory from reality but could serve as an indicator of light scattering by atmospheric particles other than molecules [1]. Then he defined turbidity depolarization (D_T) as the ratio of the intensity of the unpolarized light scattered by large solid and liquid particles suspended in the atmosphere to the total intensity of the light coming from a given point in the sky and resulting from scattering by both air molecules and large particles. After simple calculations and some simplifying assumptions, he obtained

$$D_T = 1 - \frac{p_r}{p_i} \quad \text{or} \quad p_r = (1 - D_T)\, p_i, \tag{6}$$

where p_r and p_i are the degrees of polarization in a real and ideal atmosphere (observed and theoretical), respectively. It follows from Sivkov's formula that the observed polarization is proportional to the theoretical. E. V. Pyaskovskaya-Fesenkova [14] arrived at precisely this conclusion by an independent comparison of observed polarizations and calculations by the Rayleigh and Cabannes—Tikhanovskii formulas. With the latter, for example,

$$p = p_m p_i = \frac{p_m \sin^2 \theta}{1 + k \cos^2 \theta}, \tag{7}$$

where $k = (1 - a)/(1 + a) \cong 0.92$.

Here, the proportionality factor is the maximum polarization, which, according to E. V. Pyaskovskaya-Fesenkova's observations, is a reliable measure of atmospheric transmittance [15].

Formulas (3) and (4) make it easy to incorporate into formula (1) the polarization parameter for transmittance p_m, and we obtain

$$p = \frac{p_m \sin^4 \theta}{1 - p_m \cos^4 \theta}. \tag{8}$$

This formula and E. V. Pyaskovskaya-Fesenkova's give similar values. As is shown below, the best agreement occurs with scattering angles close to 90°. The maximum discrepancy is about 3% and occurs at scattering angles close to 40 and 140°.

Scattering angles	20 and 160°	40 and 140°	50 and 130°	60 and 120°	80 and 100°	90°
Polarization, $^0/00$,						
from formulas.... (7)	44	184	291	418	646	685
(8)	20	153	267	403	645	685
Difference Δp	24	31	24	15	1	0

Note that formula (8) has been tested on a large amount of observation material.

It must be emphasized that atmospheric polarization observations can be used not only for independent estimation of turbidity and transmittance, but also to detect and study turbidity in various directions from the observer [10-12], determine the albedo of the earth's surface [13], isolate the aerosol component of polarization [16, 17], separate the effects of absorption and scattering [8], forecast optical situations [11, 18], etc. Each of these applications is based on empirical relationships that require not so much checking and verification as refinement by statistical processing of more extensive data on uniform polarimetric observations.

As has been stated, polarization is more sensitive to scattering than are other values. Its measurement, therefore, to check any theory of atmospheric light scattering, is preferable to the measurement of the intensity or attenuation of solar radiation. In the case of secondary and higher-order scattering, taking into account only intensity without polarization not only does not simplify analysis but can produce incorrect results [1].

In order for polarimetric methods to gain the same popularity as, for example, actinometric, a great deal of attention must be given to systematic observations of polarization at a number of points, designing and mass producing polarimeters and polariscopes, compiling and disseminating comprehensive handbooks on polarimetry of the day, twilight, and night sky, and introducing in the universities elective courses in scattering and polarization of light in turbid media.

LITERATURE CITED

1. Z. Sekera, "Scattering of light in the atmosphere and diffuse sky radiation," Sci. Progress, A, Vol. 45, No. 179, § 7 and 9 (1957).
2. G. V. Rozenberg, "Light scattering in the earth's atmosphere," Usp. Fiz. Nauk, Vol. 71, No. 2, pp. 173-213 (1960).
3. P. N. Boiko and G. A. Kharitonova, "Sky polarization and atmospheric transmittance," Trudy Astrofiz. inst. Akad. Nauk Kazakh SSR, Vol. 6 (1963).
4. P. N. Boiko and G. A. Kharitonova, "Maximum daylight polarization and atmospheric transmittance," in Actinometry and Atmospheric Optics [in Russian], Izd. Akad. Nauk SSSR (1964), pp. 160-164.
5. W. Milch, "Über den Einfluss grösserer Teilchen in der Atmosphäre des Polarisationverhältnisse der Himmelslichtes, Z. Geophys., Vol. 1, No. 3 (1925).
6. K. Ya. Kondrat'ev, Actinometry [in Russian], Chapter IV, Sections 4 and 4, Leningrad (1956).

7. D. G. Stamov, Sky Polarization and Atmospheric Turbidity [in Russian], Author's Abstract of Candidate's Dissertation, Moscow (1953).

8. D. G. Stamov, "On skylight polarization as a function of the meteorological state of the earth's atmosphere," Izv. Krym. ped. inst., Vol. 21, pp. 301–312 (1955).

9. D. G. Stamov, "Experience in measuring skylight polarization for various scattering angles and various states of the earth's atmosphere," Izv. Krym. ped. inst., Vol. 29, pp. 283–320 (1957).

10. D. G. Stamov, "On the possibility of polarimetric determination of turbidity in different directions," in: Actinometry and Atmospheric Optics [in Russian], Leningrad (1961), pp. 115–124.

11. D. G. Stamov, "The meteorological aspect of skylight polarization," Trudy Vses. meteorol. soveshch., Vol. 6, p. 130 (1961).

12. D. G. Stamov, "The special value of the polarimetric method of detecting and studying atmospheric turbidity as compared with the actinometric and diaphanoscopic methods," Trudy Astrofiz. inst. Akad. Nauk Kazakh SSR, Vol. 3, pp. 163–170 (1962).

13. S. I. Sivkov, "A quantitative characteristic of light depolarization by atmospheric aerosol particles," in: Actinometry and Atmospheric Optics [in Russian], Leningrad (1961), pp. 124–132.

14. E. V. Pyaskovskaya-Fesenkova, "Some data on light polarization by the atmosphere," Doklady Akad. Nauk SSSR, Vol. 131, No. 2, pp. 297–299 (1960).

15. E. V. Pyaskovskaya-Fesenkova, "Determining atmospheric transmittance from light polarization," Doklady Akad. Nauk SSSR, Vol. 134, No. 4, pp. 813–815 (1960).

16. E. V. Pyaskovskaya-Fesenkova, "On light scattering and polarization under conditions of the Libyan Desert," Doklady Akad. Nauk SSSR, Vol. 123, No. 6, pp. 1006–1009 (1958).

17. T. P. Toropova, "On the degree of polarization of light scattered in the atmospheric boundary layer as a function of wavelength," Izv. Astrofiz. inst. Akad. Nauk Kazakh SSR, Vol. 11, pp. 105–110 (1961).

18. N. N. Nikitinskaya, "On optical methods for studying atmospheric variability," in: Actinometry and Atmospheric Optics [in Russian], Izd. Akad. Nauk SSSR (1964), pp. 202–203.

SOME RESULTS OF PHOTOELECTRIC MEASUREMENTS OF ATMOSPHERIC POLARIZATION IN THE TOWN OF SHAKHTY

G. S. Isaev

Studies of the relationship between the polarization and turbidity of the atmosphere [1-3] have shown that turbidity can be detected and studied in various directions by almucantar measurements of the polarization of light scattered by the atmosphere. The almucantar measurements were made visually with a Martens polarimeter without filters and, therefore, had the following shortcomings: the results depended upon the sensitivity of the observer's eye; the observations required a great deal of time (about 30 min for one almucantar); and the measurements did not allow the total intensity or the polarized and unpolarized components to be determined simultaneously with polarization.

For an independent and objective check of the results obtained by D. G. Stamov, a semiautomatic photoelectric polarimeter was built at the Shakhty Pedagogical Institute for almucantar measurements of the maximum and minimum components of partially polarized light and its degree of polarization. With the aid of a semiautomatic device, the optical axis of the polarimeter was aimed at points on the selected almucantar at distances of 20° from one another. Polarization was measured at 17 points on each almucantar. The light passed through a uniformly rotating polaroid and entered an FEU-19M photomultiplier. The photomultiplier current was amplified by an electrometric amplifier and sent to an EPP-09 electronic recording potentiometer, which drew a curve of the photocurrent intensity as a function of the position of the axis of the instrument relative to the sun and the position of the plane of transmission of the polaroid. Polarization was measured for almucantars with zenith distances of 80, 75, 70, 65, 60, 50, 40, 30, and 20°. The measurement time for all almucantars was 45 min. Since the apparatus was only a first experimental model, it was impossible to show exactly the measurement errors. But judging from the results, we can assume that the error was not more than a few percent. For this reason, we had to average the results of a number of measurements, so the regularities we detected reflect only the general nature of the phenomenon.

Using the obtained data, we calculated the degree of polarization and plotted it as a function of the difference between the azimuths of the sun and the optical axis of the instrument (ΔA). Figure 1 shows such graphs for a number of almucantars. The graphs were plotted from averaged data for five measurements, on 26 and 27 July and 7, 10, and 20 August, 1965, at sun elevations of from 29 to 36°. The maximum polarization varied from 35 to 42%. This low polarization may be attributed to the fact that the measurements were made in the center of the town, which was surrounded by smoking waste heaps, which greatly contaminated the atmosphere. The rather good symmetry of the degree of polarization relative to the solar ver-

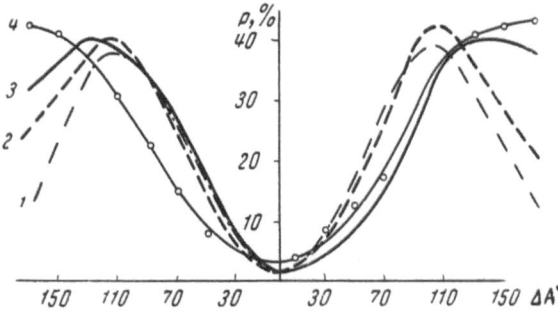

Fig. 1. Degree of polarization versus difference between azimuth of optical axis of instrument and azimuth of sun for almucantars: 1) 80°; 2) 70°; 3) 60°; 4) 40°.

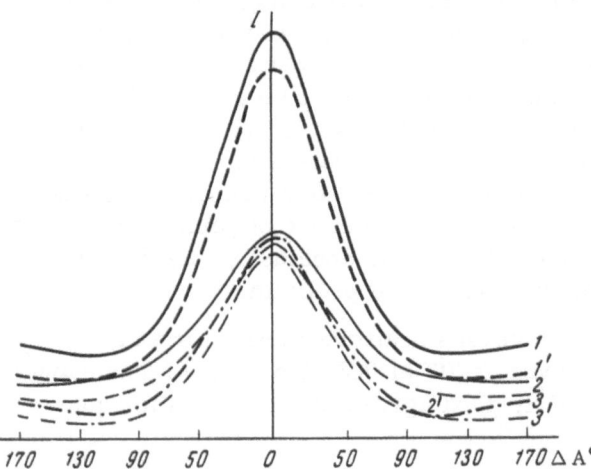

Fig. 2. Total intensity (1) and maximum (2) and minimum (3) components of partially polarized light versus difference between azimuths for almucantars: 1, 2, 3) 75°; 1', 2', 3') 60°.

tical circle may be explained by the fact that the results of five days of observations were averaged, since on some days the symmetry was sometimes destroyed.

The total intensity and the intensity components of light partially polarized in mutually perpendicular planes as functions of the distance from the observation point to the solar vertical circle were studied simultaneously. Figure 2 shows graphs of the total intensity and the maximum and minimum components as functions of the azimuth difference. The data are averages from the same five measurements. It is evident that the overall shape of the curves is approximately the same. They have a maximum at $\Delta A = 0°$, toward the sun, and a less-pronounced second maximum in the antisolar direction. The minima of the curves are less distinct. The graphs of the maximum and minimum components as functions of ΔA are the ones of interest here, since the total intensity is equal to their sum. The minimum component has a more pronounced decay and a more clearly expressed minimum, which is within 100–120°. This minimum is easy to see at low almucantars with zenith distances of 80, 75, and 70°; the curve smooths out as the zenith distance of the almucantar decreases. Since the graph of the minimum component duplicates that of the unpolarized light ($I_{unp} = 2I_{min}$), its variation can evidently be explained by the fact that there is a greater amount of polarized light in this range of angles, because maximum polarization is observed precisely here. As the zenith distance of the almucantar for the given solar elevations decreases and becomes smaller than the solar zenith distance, the total light intensity and, moreover, the amount of polarized light are reduced, since the degree of polarization is reduced by a change in the scattering angle. This makes the minimum in the minimum component less pronounced.

It is interesting to see that there are appreciable deviations from the expected in the overall variation of the polarized part. According to the Rayleigh theory, the variation of the polarized light must duplicate the variation of polarization, i.e., have a maximum at 90° and then decay smoothly as the scattering angle decreases to zero or increases to 180°. While the variations of the degree of polarization for light scattered by the atmosphere mostly confirm this law, there are deviations in the variation of the polarized part of the light. The maxima of the curves are, as a rule, poorly defined, do not coincide with the polarization maxima, and are shifted toward the solar vertical circle, i.e., in the direction of smaller scattering angles. This can also be seen by comparing the variation of the polarized part for Rayleigh scattering and the curve that we obtained for the polarized part as a function of the scattering angle. In my opinion, these deviations can be attributed to the fact that light polarized

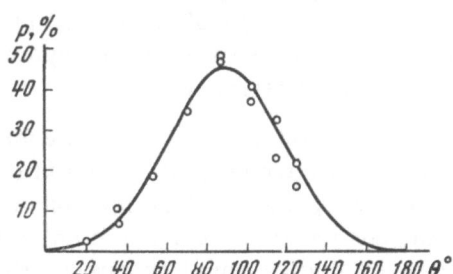
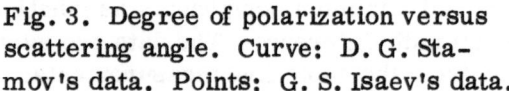
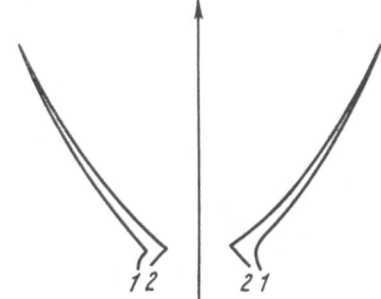

Fig. 3. Degree of polarization versus scattering angle. Curve: D. G. Stamov's data. Points: G. S. Isaev's data.

Fig. 4. Scattering indicatrices for entire atmosphere: 1) for total intensity; 2) for unpolarized portion.

by aerosol scattering is added to the light polarized by molecular scattering. The value of the aerosol-polarized component must be a function of the amount of aerosol and the scattering angle, and this relationship differs greatly from the one for molecular scattering, especially for small scattering angles. So the total amount of polarized light shows up as a combination of two types of polarized light — molecular and aerosol. And if this is true, then the amount of polarized light can serve as a good indicator of the presence of aerosols in the scattering medium. This indicator is considerably more sensitive than the degree of polarization, since an increase in total intensity at small scattering angles affects the degree of polarization much more than does an increase in the amount of polarized light. These conclusions need additional study and experimental verification and should be considered preliminary.

Figure 3 shows a graph of degree of polarization versus scattering angle for an arbitrarily selected date, 27 July 1965, with $z_\odot = 60°$ and $z_A = 70°$ at 0639 local time. The same relationship from D. G. Stamov's data, which is in fairly good agreement, is shown for comparison.

We made an attempt, on the basis of our photoelectric measurements, to construct the scattering indicatrix for the entire atmosphere. Figure 4 shows scattering indicatrices for the total intensity and the unpolarized part.

We may conclude that our instrument is suitable for studying the atmosphere, since the results it provided are in fairly good agreement with the general regularities detected by other authors, particularly D. G. Stamov. At the same time, it enables us to study the polarization state in short time intervals, which is very important in research on such a highly mutable object as the atmosphere. Work to improve the photoelectric apparatus is now going on at the Shakhty Pedogogical Institute.

I thank Professor E. V. Pyaskovskaya-Fesenkova and Docent D. G. Stamov for their advice.

LITERATURE CITED

1. D. G. Stamov, Sky Polarization and Atmospheric Turbidity [in Russian], Author's Abstract of Candidate's Dissertation, Moscow (1953).
2. D. G. Stamov, "The special value of the polarimetric method of detecting and studying atmospheric turbidity as compared with the actinometric and diaphanoscopic methods," in: Light Scattering and Polarization in the Earth's Atmosphere [in Russian], Alma-Ata (1962).

3.   D. G. Stamov, "On the possibility of polarimetric determination of turbidity in different directions," in: Actinometry and Atmospheric Optics [in Russian], Leningrad (1961).

ON STUDYING TWILIGHT PHENOMENA

V. G. Fesenkov

Twilight phenomena are very complex, since the brightness observed at any point in the sky is the sum of the brightnesses of the primary twilight segment and of higher-order scattering, to which, when the sun has sunk sufficiently below the horizon, we must add the background of the night sky — primarily zodiacal light.

Twilight can be used to study the optical properties of the higher layers of the atmosphere, whose height increases all the more as the sun sinks below the horizon, while the lower layers gradually enter the shadow zone.

As has been suggested earlier [1], to distinguish the brightness of primary twilight, one should simultaneously observe two symmetric points on the solar vertical circle at, for example, a zenith distance of 70°. Starting when the sun is not less than 5-6° below the horizon, the primary twilight segment is no longer present at the opposite symmetric point, and only higher-order, chiefly tropospheric scattering remains.

At the first symmetric point, the observed brightness is the sum of three components: primary twilight I_1, the tropospheric component I_2, and the night-sky background I_3, which is primarily zodiacal light:

$$I = I_1 + I_2 + I_3.$$

At the second symmetric point, we have only

$$I' = I_2' + I_3'.$$

Components I_3 and I_3' are found directly from observations made before the complete cessation of twilight. If we determine the ratio of the components

$$K = \frac{I_2}{I_2'},$$

we quickly find the brightness of the primary twilight segment

$$I_1 = I - I_3 - KI_2'.$$

We made calculations on the basis of this schematic distribution of brightness in the primary twilight segment under the assumption that twilight was produced by only one effective ray 20 or 30 km above the earth's surface (30 km, if ozone is taken into account). If, moreover, each element of the primary twilight segment is considered an unpolarized external light source, then for the desired ratio K we obtain

$$K = \frac{\iint B(z, A) f(\vartheta_1) \varphi(z, z_1) \sin z \, dz \, dA}{\iint B(z, A) f(\vartheta_2) \varphi(z, z_1) \sin z \, dz \, dA},$$ (1)

where $B(z, A)$ is the brightness of element $d\sigma = \sin z \, dz \, dA$, and $f(\vartheta_1)$, $f(\vartheta_2)$ are the corresponding scattering indicatrices, which are found from daytime brightness observations. In the case of only first-order scattering, the function $\varphi(z, z_1)$ has the usual expression

$$\varphi(z, z_1) = \frac{p^{\sec z} - p^{\sec z_1}}{\sec z_1 - \sec z} \sec z_1,$$

where p is the atmospheric transmittance and z_1 is the zenith distance of the symmetric points.

For a more rigorous calculation of the factor K, we must proceed from the general expression for the brightness of primary twilight

$$B(z, A) = L \frac{f^0(\vartheta)}{\sin \vartheta} \int_{h_0 \min}^{\infty} \mu(h) (abs) \, dh_0,$$

where each trajectory of the solar light ray has its own minimum distance h_0 from the earth's surface and corresponding extinction (abs) over its entire extent. The function $f^0(\vartheta)$ is the atmospheric scattering function, which depends upon h, which, in turn, is related to h_0 as follows:

$$h = \frac{\sin z - \sin(z - \alpha)}{\sin(z - \alpha)}, \quad \tan \alpha = \frac{\tan z (1 + h_0 - \sin \zeta)}{1 + h_0 - \cos \zeta \tan z},$$

where z is the zenith distance of the observed point of the primary twilight segment; ζ the solar zenith distance; and α an auxiliary angle. The angle ϑ is the angle between the incident and scattered rays in each illuminated element of the atmosphere on the line of sight of the observer.

If we take into account polarization of the primary twilight segment, we obtain for the desired factor K the expression

$$K = \frac{\iint B(z, A) f(\vartheta_1) \varphi(z, z_1) (1 + P_1 P_0 \cos 2\alpha_1) \, d\sigma}{\iint B(z, A) f(\vartheta_2) \varphi(z, z_2) (1 + P_2 P_0 \cos 2\alpha_2) \, d\sigma}.$$ (2)

The total scattering indicatrix $f(\vartheta)$ consists of two components f' and f'', which are polarized in mutually perpendicular directions relative to the great circle, which passes through the given symmetric point on the solar vertical circle and the element of the primary twilight segment under examination. Thus,

$$f = f' + f''.$$

In expression (2),

$$P_1 = \frac{f_1' - f_1''}{f_1' + f_1''}$$

and

$$P_2 = \frac{f_2' - f_2''}{f_2' + f_2''}$$

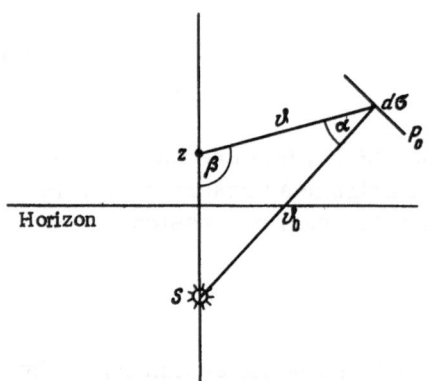

Fig. 1. First symmetric point on solar vertical circle relative to sun S and element $d\sigma$ of primary twilight segment.

are the polarizations at the symmetric points on the solar vertical circle; P_0 is the polarization of element $d\sigma$ = sin zdzdA of the primary twilight segment; and α_1 and α_2 are the angles at element $d\sigma$ formed by the directions from points 1 and 2, respectively, on the solar vertical circle and from the sun. Figure 1 shows the first symmetric point relative to the sun S and element $d\sigma$.

It is very difficult to determine the degree of polarization P_0 at various points on the primary twilight segment. In essence, for this we need to know the scattering indicatrix for the higher atmospheric layers, which are rich in aerosols of cosmic origin. In all probability, this indicatrix has an extremely irregular structure. Such determinations could be made with high-altitude balloons, but, for the time being, we are forced to use the usual indicatrix obtained from clear-sky, dry-weather observations, when the sun is in a low position, when the equivalent atmospheric layer is raised to a sufficient height.

Then the aerosol component is extracted from this indicatrix [2, 3]. The results of such a determination of both the total indicatrix f and its aerosol component f^0, as well as their corresponding degrees of polarization P and P_0 are given below:

ϑ	f	P	f^0	P_0	ϑ	f	P	f^0	P_0
10°	4.22	0.020	10.13	0.020	100°	1.000	0.715	0.934	0.425
15	3.183	0.035	6.60	0.040	110	1.063	0.633	0.890	0.405
20	2.675	0.054	4.95	0.062	120	1.177	0.520	0.894	0.380
30	2.075	0.125	3.153	0.120	130	1.266	0.383	0.898	0.335
40	1.707	0.220	2.235	0.180	140	1.362	0.270	0.900	0.280
50	1.452	0.333	1.733	0.240	150	1.432	0.130	0.894	0.222
60	1.279	0.475	1.408	0.308	160	1.496	0.050	0.890	0.153
70	1.152	0.607	1.246	0.360	170	1.540	0.010	0.890	0.075
80	1.046	0.710	1.333	0.405	180	1.567	0.000	0.890	0.000
90	1.000	0.750	1.000	0.424					

Calculations of K by formulas (1) and (2) are very tedious, since they require preliminary calculation of the primary twilight segment.

For a solar zenith distance of 100°, formulas (1) and (2) give K = 2.388 and K = 2.301.

Thus, in correcting the observed brightness I for I_3, in order to obtain the brightness of the primary segment I_1 we must subtract from I not simply I_2' but the considerably larger value KI_2'. The difference is very small and has real significance only when the optical properties of the atmosphere and the underlying surface are completely uniform.

Now we must consider the following limitations of the method. In deriving formulas (1) and (2) it was assumed that each element of the primary twilight segment could be considered an external light source, even if partially polarized. This is valid only for a sufficiently distant region of the atmosphere illuminated by direct solar rays, which can occur only when the sun is not less than 6° below the horizon, i.e., after civil twilight. On the other hand, when the sun has sunk more than 10° below the horizon, the point on the solar vertical circle z = 70° at which observations are being made is already in a very weak part of the primary twilight segment. In this case, the brightness of the observed point is almost entirely a result of tropospheric scattering and is therefore almost completely eliminated when K is calculated properly.

The proposed method has a limited area of application and, moreover, requires knowledge of the real scattering indicatrix for high atmospheric layers. But this property of the

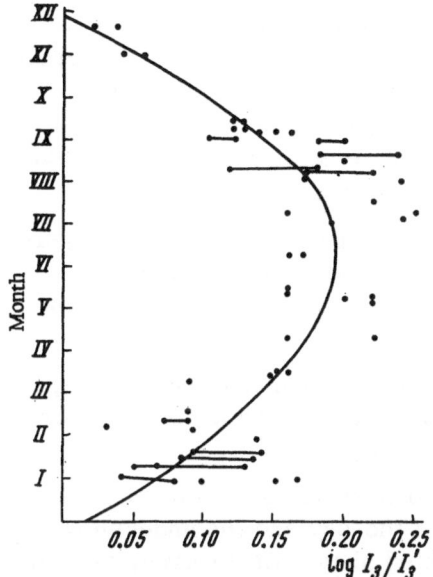

Fig. 2. Residual difference between intensities at symmetric points on solar vertical circle at z = 70° after cessation of atmospheric twilight for various months.

method allows us, by using it for the maximum possible solar zenith distance, to find the factor K from observations alone. In fact, then we have

$$I = I_2 + I_3 \text{ and } I' = I_2' + I_3',$$

hence,

$$K = \frac{I - I_3}{I' - I_3'} = \frac{I_2}{I_2'},$$

since I_1 is a negligible in comparison with the observed value I. If we find K from observations, we can represent it theoretically by selecting the proper scattering indicatrix. The K values calculated from the thus-corrected indicatrix can serve as a reliable correction of the brighter parts of the observed twilight curve for the effect of tropospheric scattering. This allows us to reliably isolate the primary twilight component for a given small interval $\Delta \zeta$. The remainder of the twilight curve, which corresponds to the period up to the end of twilight, represents essentially only one tropospheric component and there is no point in observing it.

Thus, the main shortcoming of this method is that the desired value I_1 is found as the difference between two comparable values I and I_2 and can be negligible. Its application is very limited and is essentially the solar zenith distance range of 4-5°. The situation is not improved when the observed point is in a lower and, therefore, brighter part of the twilight segment, since this reduces considerably the effective height.

Systematic twilight observations have been made by this method for several years at the Mountain Astrophysical Observatory (Kamenskoe Plateau, Alma-Ata) and at Abastuman. A general comparison of the obtained twilight curves reveals the following distinctive features.

With the approach of night, when the twilight curve begins to run practically parallel to the axis of the abscissas, the brightness at the second symmetric point is usually less than at the first. The difference between the brightnesses depends upon the time of year, as can be seen from our comparison (Fig. 2) of the material from Kamenskoe Plateau. Immediately after twilight, the night sky on the side facing the sun is still considerably brighter than in the opposite direction, especially in summer and in the beginning of autumn. In all probability, this is due to the varying incidence of the ecliptic to the horizon, i.e., the varying effect of zodiacal light. If this is true, then it should evidently be concluded that the dust cloud around the earth makes a negligible contribution to the over-all phenomenon of deep twilight. In fact, the cloud encircles the entire earth more or less uniformly, particularly in its lower layers, so it should always make a fairly large difference in brightness between given directions. But in reality, the observed difference indicates that the dust is concentrated primarily in the plane of the ecliptic. Comparison of the twilight curves for Abastuman and for Kamenskoe Plateau shows that the former have a considerably more pronounced decay than the latter.

Generally speaking, the twilight curves for both points are stable enough that we can draw their typical average shape. But if we compare the curves obtain at Kamenskoe Plateau among themselves, we see that they differ in many details. On 4 October 1962, for example, the distance between the twilight curves at both symmetric points in the blue rays was at first considerably greater than usual, and then, starting with a solar zenith distance of 99°, it rapidly decreased and approached normal. In the yellow rays, the curves remained normal in general. This interesting characteristic shows clearly that on that night at a height of about 90 km there was something like noctilucent clouds, i.e., some kind of layer of particles that scattered mainly rays of short wavelength but was transparent to yellow rays.

In general, the shape of twilight curves is almost independent of wavelength, but the yellow (560 μ) brightness of the night sky is considerably greater than the blue (440 μ). Sometimes there is an overall reduction in twilight brightness, and sometimes an overall increase, without a change in the curve shape.

All of this shows that we can trace even fine turbidity in high atmospheric layers due, for example, to fine cosmic dust and its gradual settling into lower layers. This could be very important in determining the rate of precipitation of cosmic material.

A shortcoming of this twilight method, which is based on observations of only twilight, even if at different wavelengths, is that the optical properties of the sky and the underlying surface must be completely homogeneous. And, as has been noted above, its accuracy is comparatively low. Such observations should therefore be made, whenever possible during expeditions, in the most suitable localities.

The simultaneous determination of two Stokes parameters, i.e., brightness and degree of polarization, offers great possibilities.

At a point on the solar vertical circle for comparatively bright twilight we have

$$I_1 = I \frac{P - P_2}{P_1 - P_2},$$

and for weaker twilight, when zodiacal light and the background of the night sky have already begun to exert an influence,

$$I_1 = \frac{I(P - P_2) + I_3(P_2 - P_3)}{P_1 - P_2}.$$

where I and P are the directly observed intensity and degree of polarization at the point on the sun's side of the vertical circle; P_1 is the degree of polarization of the primary twilight segment, which is found directly as a function of the angular distance from the sun, on the basis of the scattering indicatrix with its polarization components; and P_3 is the polarization of zodiacal light, which is also found from night observations. This method is discussed in greater detail in [2].

For P_2 we have

$$P_2 = \frac{\iint B(z, A)(f' + f'') \varphi(z, z_1) \cos 2\beta (P + P_0 \cos 2\alpha) \, d\sigma}{\iint B(z, A)(f' + f'') \varphi(z, z_1)(1 + PP_0 \cos 2\alpha) \, d\sigma},$$

where β is the angle between the sun and element $d\sigma$ at the point in question on the solar vertical circle. The remaining symbols are explained above (see Fig. 1).

Theoretically, it is sufficient to make observations at only one point in the sky toward the sun, so this method is independent of the inevitable heterogeneity of the optical properties of the atmosphere at different azimuths. But at the second symmetric point, where only tropospheric scattering is present, P_2' is found directly from observations, and this lets one check the calculations for P_2. An example calculation for $\zeta = 100°$ and $z = 70°$ for the first symmetric point gave the very small value $P_2 = 0.0488$.

The simultaneous determination of three Stokes parameters — brightness, degree of polarization, and its orientation — offer even greater possibilities. Since in the solar vertical circle the angle of polarization is always perpendicular to the vertical circle, observations should be made to the side of it, for example, at a constant difference between the azimuths of the given point and the sun and comparatively close to the horizon, since local conditions allow this.

Test observations of this kind were made by the Astrophysics Institute of the Academy of Sciences of the Kazakh SSR (at Kamenskoe Plateau) in 1959 and 1960 for brightness and polarization at the celestial pole of the day and twilight sky. Certain regularities manifested themselves quite definitely. The orientation of the predominant vibrations at the celestial pole showed a regular linear variation as a function of the hour angle during the day. After sunset, the variation of the angle of polarization at first remained as before, to a solar zenith distance of approximately 94-95°, and then it began to change more and more slowly.

The main characteristic of the degree of polarization was that, in spite of slight fluctuations during the day according to the overall atmospheric transmittance (the more atmospheric haze, the less polarization), as the sun approached the horizon this value increased rather rapidly and reached approximately the same limiting values regardless of atmospheric transmittance. This was due to the fact that the effective layer, which chiefly determines the optical properties of the atmosphere, when the sun is in a low position is separated from the earth's surface and lifted extremely high, even beyond the troposphere. The increase in the degree of polarization at the celestial pole continues even after sunset, reaches a maximum when the solar zenith distance ζ is about 93° and then begins to decrease, at first slowly and then more rapidly (when $\zeta > 96°$), becoming completely negligible at deeper twilight. The brightness of the sky decreases continuously and finally becomes the overall luminosity of the night sky (Fig. 3).

This variation in the angle α and the degree of polarization P is easily interpreted if we consider the interaction of the primary twilight segment and the tropospheric component, taking into account that the former, which is characterized by only first-order scattering and, therefore, by an entirely regular orientation of polarization, predominates at first. Then the slightly polarized tropospheric component, which depends upon the illumination of the entire sky by primary twilight, the positions of whose isophots are practically unchanged relative to the horizon, begins to assert itself to a greater and greater extent.

Let us assume that in the primary twilight segment the intensity distribution and the polarization, which is determined by the position of the sun below the horizon, are known from preliminary calculations. We take a point outside the solar vertical circle and hold it, for example, at a constant azimuth from the sun at the same height above the horizon. At this point we determine the three values I, P, and α for a particular moment in time. The direction of twilight polarization is determined by the direction MS to the sun, and the value P_1 follows from the adopted scattering indicatrix, which has been separated into its respective polarizing components. On the other hand, if we proceed from the necessary condition that the scattered radiation produced by some partially polarized element $d\sigma$ of the primary twilight segment is incoherent, then the corresponding vector $I_2 P_2$ must be perpendicular to the arc of the great circle (Mdσ) and must make with the vertical circle at point M an angle β, which can be determined.

Consider the following problem. Let the partially polarized light from two sources — the sun and the primary twilight segment — be superimposed at some point M outside the solar vertical circle. The corresponding intensities are

$$I_1 = a_1^2 + b_1^2; \qquad I_2 = a_2^2 + b_2^2$$

(a and b are the maximum and minimum vibrations in mutually perpendicular directions).

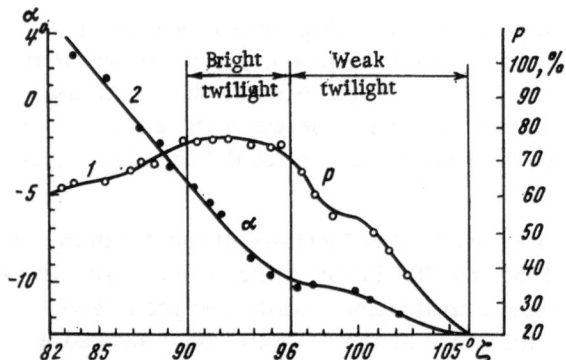

Fig. 3. Degree of polarization P (curve 1) and angle of polarization α (curve 2) at celestial pole, versus solar zenith distance ζ during transition from day to deep twilight.

If we let P_1 and P_2 be the degrees of polarization, we have

$$I_1 P_1 = a_1^2 - b_1^2; \qquad I_2 P_2 = a_2^2 - b_2^2 .$$

Let the vectors $I_1 P_1$ and $I_2 P_2$ form the angle β, and let the resultant vector IP form with $I_1 P_1$ the angle α. If we project vibrations a_1, b_1, a_2, and b_2 onto IP and onto the perpendicular to it, we find

$$a^2 = (a_1^2 - b_1^2) \cos^2 \alpha + (a_2^2 - b_2^2) \cos^2 (\beta - \alpha);$$

$$b^2 = (a_1^2 - b_1^2) \sin^2 \alpha + (a_2^2 - b_2^2) \sin^2 (\beta - \alpha);$$

therefore,

$$IP = I_1 P_1 \cos 2\alpha + I_2 P_2 \cos (2\beta - 2\alpha).$$

Obviously, the desired angle α must satisfy the condition

$$\frac{d(IP)}{d\alpha} = 0; \qquad \frac{d^2(IP)}{d^2\alpha} < 0.$$

Hence, we find that

$$\tan 2\alpha = \frac{\sin 2\beta}{\dfrac{a_1^2 - b_1^2}{a_2^2 - b_2^2} + \cos 2\beta}$$

or

$$\tan 2\alpha = \frac{I_2 P_2 \sin 2\beta}{I_1 P_1 + I_2 P_2 \cos 2\beta} .$$

Substitution of this value of α into the above expression for IP gives, after transformations,

$$IP = \sqrt{(I_1 P_1)^2 + (I_2 P_2)^2 + 2(I_1 P_1)(I_2 P_2) \cos 2\beta}.$$

Actually, the inverse problem must be solved. From observations we determine at point M the second Stokes parameter IP, and also its sense, i.e., the angle which the polarized component makes with vector $I_1 P_1$, which is perpendicular to the direction to the sun.

We also assume that the sense of $I_2 P_2$ is known, since it is determined by the position of the bright regions of the primary twilight segment near the solar vertical circle. It is necessary to isolate the vector $I_1 P_1$ and with it I_1, i.e., the brightness of the primary twilight segment at the point in question.

As before, let α be the angle that IP makes with $I_1 P_1$ and let β be the angle between vectors $I_1 P_1$ and $I_2 P_2$. Then we obtain

$$I_1 P_1 = IP \frac{\sin(2\beta - 2\alpha)}{\sin 2\beta} .$$

Thus, to determine the brightness of the primary twilight segment at a point M outside the solar vertical circle, it is sufficient to find I, P, and α at this point and then, if we know the angle β between the direction to the sun and the direction to the effective region of the primary twilight segment, we immediately find $I_1 P_1$. The degree of polarization P_1 of primary twilight is found from the standard scattering indicatrix for the upper atmosphere.

In order to determine the angle β more accurately, we must discuss the problem of determining the vector $I_2 P_2$ for the selected point M.

In the particular case when M is on the solar vertical circle, the preferred direction of vibration is normal to this vertical circle, wherein, as was indicated above,

$$I_2 P_2 = \iint B(z, A)(f' + f'') \varphi(z, z_1) \cos 2\beta (P + P_0 \cos 2\alpha) \, d\sigma.$$

If point M is not on the solar vertical circle, this expression remains valid if β is taken to mean the angle at M between the direction to element $d\sigma$ and the direction to the effective region N of the primary twilight segment. This direction is characterized by the maximum possible value of P_2 for any orientation of line MN.

This method makes it possible to find only the second Stokes parameter for primary twilight. In order to obtain the intensity itself I_1, which directly characterizes the scattering properties of the upper atmospheric layers, we must also know the exact value of the degree of polarization P_1 due to high aerosols, which, in all probability, are of cosmic origin. This gives particular practical importance to determination of the nature of the scattering indicatrix from high-altitude balloons. But in the mean time, we must use the scattering indicatrix as determined from observations of the daytime sky, with the aerosol component eliminated from it.

LITERATURE CITED

1. V. G. Fesenkov, Trudy Astrofiz. inst. Akad. Nauk Kazakh SSR, Vol. 3, p. 214 (1962).
2. E. V. Pyaskovskaya-Fesenkova, Investigation of Light Scattering in the Earth's Atmosphere [in Russian], Izd. Akad. Nauk SSSR, Moscow (1957), p. 106.
3. E. V. Pyaskovskaya-Fesenkova, in: Light Scattering and Polarization in the Earth's Atmosphere [in Russian], Izd. Akad. Nauk Kazakh SSR (1962), p. 123.

THE HEIGHT OF THE TWILIGHT RAY

N. B. Divari

Scattering of direct solar radiation by the earth's atmosphere under twilight conditions occurs in a certain effective layer, whose height is a function of the sinking of the sun. This was first demonstrated by V. G. Fesenkov [1] and later examined in detail by N. M. Shtaude [2]. The thickness of the layer that determines the main flow of scattered radiation is not great, so in the first approximation the layer is often replaced by a single ray (half-line), which N. M. Shtaude called the "twilight ray." This twilight ray determines the mean height of the scattering layer for a given solar zenith distance. In Fig. 1, the twilight ray is shown by line SM, which is at distance $R + h_0$ from the center of the earth (R is the radius of the earth). When observations are made in direction BM, the height of the twilight layer in this direction is the difference $h = OM - R$, i.e., the height above the earth's surface of the point of intersection of the line of sight BM and the twilight ray SM.

The fact that the distance of the twilight ray from the earth's surface is given as almost constant has been considered the most interesting feature of this representation of light scattering under twilight conditions. As N. M. Shtaude has shown [2], for observations at the zenith, height h_0 is practically unchanged for distances of 1 to 14° of the sun below the horizon, and it can be taken as 20 km. Later, however, Shtaude [3] pointed out that h_0 must be a function of the sun's position, but she did not give a relationship for this. In [4], we indicated that h_0 is also a function of the wavelength of the monochromatic radiation at which the observations are made or, what is the same, of the transmittance of the atmosphere.

Since determinations of the height of the point M of maximum scattering of direct solar radiation are of great value in the study of the optical properties of the atmosphere, we examined the dependence of height h upon the sinking of the sun g and the transmittance of the atmosphere (or the wavelength of the scattered light). This was done assuming that the solar rays were rectilinear (ignoring refraction) and without taking into account the finite dimensions of the solar disk.

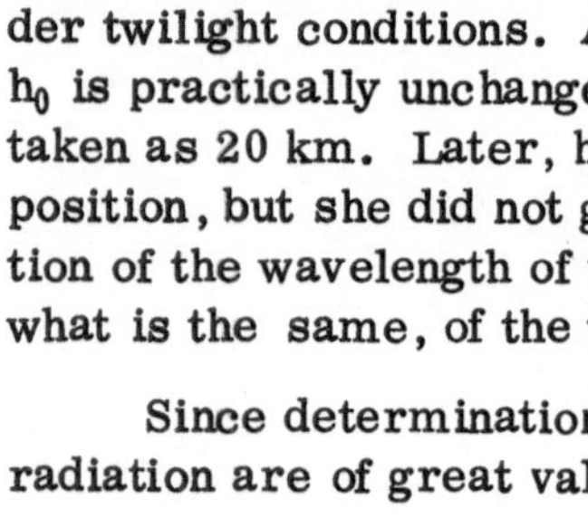

If we consider the entire sky rather than merely points on the solar vertical circle, then instead of a twilight ray we have a twilight cylinder whose generatrices are parallel to the direction of the solar rays and whose directrix is a circle lying in a plane perpendicular to the direction of the solar rays. The radius of this circle is $R + h_0$ and its center is at the center of the earth.

Let us use a rectangular coordinate system with the origin at the center of the earth, the OZ axis directed at the zenith of the

Fig. 1. Geometry of rays.

observer, and the OY axis coincident with the line of intersection of the planes of the solar vertical circle and the horizon of the observer. Then the equation of the twilight cylinder can be represented as

$$x^2 + (y \sin g + z \cos g)^2 = (R + h_0)^2. \tag{1}$$

For the equations of the line of sight BM, which is determined by the zenith distance ζ and the azimuth A as read from the solar vertical circle, we have

$$x = y \tan A; \quad z = y \cot \zeta \sec A + R \quad (A \neq 90°). \tag{2}$$

If we solve Eqs. (1) and (2) jointly, we obtain the coordinates of the point M of intersection of the line of sight with the twilight cylinder. Height h is defined as

$$h = \sqrt{x^2 + y^2 + z^2} - R. \tag{3}$$

The coordinates x, z can be determined from (2) (A $\neq$ 90°) if y is found from

$$y^2 (\tan^2 A + \sin^2 g + \sin 2g \cot \zeta \sec A + \cot^2 \zeta \sec^2 A \cos^2 g) + \\ + y (R \sin 2g + 2R \cot \zeta \sec A \cos^2 g) + R^2 \cos^2 g - (R + h_0)^2 = 0. \tag{4}$$

When A = 90°,

$$y = 0; \qquad z = x \cot \zeta + R, \tag{5}$$

where x is given by

$$x^2 (1 + \cot^2 \zeta \cos^2 g) + 2Rx \cot \zeta \cos^2 g + R^2 \cos^2 g - (R + h_0)^2 = 0. \tag{6}$$

These formulas enable us to determine the height h of the scattering layer for any point (ζ, A) in the sky if we know h_0 and g. It is also easy to derive standard working formulas for the height h_0 of the twilight cylinder (ray) from the known, for a given point (ζ, A), height h of the twilight layer and sun position g:

$$h_0 = \sqrt{x^2 + (y \sin g + z \cos g)^2} - R. \tag{7}$$

When A $\neq$ 90°, coordinates x, z are determined by formulas (2) and y is found from

$$y^2 + yR \sin 2\zeta \cos A - (h^2 + 2Rh) \cos^2 A \sin^2 \zeta = 0. \ (y \sec A > 0). \tag{8}$$

When A = 90°, we have

$$\begin{aligned} x &= \sin \zeta [- R \cos \zeta + \sqrt{(R + h)^2 - R^2 \sin^2 \zeta}], \\ y &= 0, \\ z &= x \cot \zeta + R. \end{aligned} \tag{9}$$

Height h can be obtained by calculating the brightness of primary twilight. In [5], we described calculations of the brightness $B_M(\zeta, A)$ of primary twilight, which is proportional to the integral over the line of sight BM:

$$B_M(\zeta, A) \sim \int IFn(h) \, du. \tag{10}$$

where n(h) is the number of air molecules per unit volume at height h above the earth's surface; I the solar radiation flux on the line of sight BM; F a factor that takes into account absorption of scattered radiation on the path from the scattering element to the observer; and du is an element of length of line BM.

The calculations were made with a Ural-2 computer, which printed out the values of the integrand IFn(h), from which we determined graphically its maximum, which represented the

TABLE 1. Mean Values of h_0 for Two Sun Positions

Zenith distance of point	Azimuth						
	0°	20°	40°	60°	90°	135°	180°
g = 2°							
0°	11.1						
20	11.3				11.1		10.9
40	11.5	11.5	11.4	11.3	11.1	11.8	11.6
60	10.9	10.9	10.7	10.5	11.1	11.4	11.1
70	11.3	11.3	11.1	10.7	11.1	11.0	11.3
80	12.3	12.2	11.9	11.3	11.1	10.5	9.2
g = 16°							
0	36.8						
20	37.0				36.9		36.4
40	35.9	37.5	34.4	35.4	37.2	38.6	39.4
60	31.6	29.6	29.9	33.9	37.2	36.6	36.9
70	28.5	27.3	25.5	32.8	36.8	37.3	—
80	22.8	25.8	23.3	29.9	35.4	44.7	—

TABLE 2. Smoothed Values of h_0, km

g	$\lambda = 0.37\,\mu$ $p = 0.59$	$\lambda = 0.50\,\mu$ $p = 0.85$	$\lambda = 0.58\,\mu$ $p = 0.89$	$\lambda = 0.70\,\mu$ $p = 0.95$
2°	24.8	16.4	11.2	6.0
4	25.7	17.1	11.8	6.2
6	26.5	17.8	12.4	6.4
8	27.8	18.8	13.4	7.2
10	29.3	20.4	15.0	8.9
12	31.8	24.0	19.4	13.4
14	35.0	28.8	25.1	20.0
16	38.6	35.4	33.7	31.0
18	44.3			

height of maximum scattering of solar radiation, i.e., height h. The h values were found for 32 points with $\zeta \le 80°$, sun positions of 2, 4, 6, 10, 14, and 16°, and for monochromatic radiation of 0.37, 0.50, 0.58, and 0.70 μ, which corresponded to transmission coefficients p = 0.59, 0.85, 0.89, and 0.95. From these h values, using formulas (7)-(9), we found h_0 for each of the 32 points. It turned out that for a given sun position, h_0 can with reasonable accuracy be considered constant for all points with $\zeta \le 80°$. In order to show which h_0 values were obtained for various points at the same g value, Table 1 gives as an example for λ = 0.58 μ the h_0 values in kilometers obtained for the two extreme sun positions g = 2° and g = 16°.

These data gave the following mean values of h_0: 11.2 ± 0.1 km for g = 2° and 33.7 ± 0.09 km for g = 16°. Mean values of h_0 were also obtained for the other sun positions and wavelengths, and they are given in Table 2. (The values for g = 8° and g = 12° were obtained by interpolation.)

Figure 2 shows h_0 as a function of g for the four wavelengths, and Fig. 3 gives h_0 as a function of wavelength for the sun positions 2-18°. As can be seen, h_0 is a monotonic increasing function of g for all of the wavelengths examined. At small g, h_0 increases slowly, but at g > 10°, the increase is rather abrupt. Height h_0 decreases with wavelength, but the difference between the h_0 values decreases as g increases, so that h_0 is practically same for all wavelengths when g = 18° and is 44 km (obtained by extrapolation). If g is increased further, the curves in Fig. 2

TABLE 3. Values of h (km) Calculated from h_0 Values in Table 2

Zenith distance of point	Azimuth						
	0°	20°	40°	60°	90°	135°	180°

$$\lambda = 0.37 \; \mu \; ; \; p = 0.59$$

$g = 2°$

0°	28.7						
20	28.3	28.4	28.4	28.5	28.7	29.0	29.1
40	27.9	28.0	28.1	28.3	28.7	29.3	29.6
60	27.2	27.3	27.5	27.9	28.7	30.1	30.8
70	26.6	26.7	27.0	27.5	28.7	31.0	32.3
80	25.4	25.5	25.9	26.6	28.7	34.4	39.0

$g = 4°$

0	41.3						
20	40.3	40.4	40.5	40.8	41.3	42.1	42.4
40	39.1	39.2	39.6	40.2	41.3	43.2	44.0
60	37.2	37.4	38.0	39.1	41.3	45.4	47.5
70	35.4	35.6	36.5	38.0	41.3	48.4	52.7
80	31.8	32.2	33.3	35.5	41.3	59.5	80.2

$g = 6°$

0	61.7						
20	59.5	59.6	60.0	60.6	61.7	63.5	64.2
40	56.9	57.2	57.9	59.2	61.7	65.9	68.0
60	52.9	53.3	54.6	56.8	61.7	71.3	76.8
70	49.3	49.8	51.5	54.5	61.7	78.8	91.2
80	42.4	43.1	45.4	49.6	61.7	109.1	210.2

$g = 8°$

0	90.7						
20	86.3	86.6	87.3	88.4	90.7	94.1	95.6
40	81.5	82.0	83.4	85.7	90.7	99.2	103.4
60	74.2	74.9	77.3	81.3	90.6	110.6	123.2
70	67.9	68.9	71.9	77.2	90.6	127.2	160.3
80	56.8	58.0	61.8	68.9	90.4	198.1	902.4

$g = 10°$

0	128.0						
20	120.4	120.8	122.1	124.1	128.0	134.2	137.0
40	112.2	113.5	115.4	119.4	128.0	143.4	151.5
60	100.3	101.5	105.4	112.0	127.9	165.3	192.1
70	90.6	92.1	96.8	105.2	127.8	198.6	283.3
80	74.2	76.0	81.7	92.7	127.2	340.7	

$g = 12°$

0	174.8						
20	162.5	163.2	165.2	168.4	174.8	185.1	189.8
40	149.5	150.8	154.6	160.8	174.7	200.9	215.2
60	131.5	133.4	139.2	149.3	174.5	239.7	293.9
70	117.5	119.6	126.5	139.2	174.2	301.5	517.7
80	94.9	97.4	105.4	121.3	172.9	548.7	$3.6 \cdot 10^5$

$g = 14°$

0°	231.1						
20	212.3	213.3	216.3	221.2	231.0	247.2	254.8
40	193.1	194.9	200.5	209.8	230.9	272.6	296.7
60	167.4	170.0	178.2	192.8	230.9	337.6	443.2
70	148.0	151.1	160.6	178.4	229.6	443.6	1014
80	118.5	121.8	132.7	154.3	227.1	822.4	$1.7 \cdot 10^5$

Table 3 (cont.)

Zenith distance of point	Azimuth						
	0°	20°	40°	60°	90°	135°	180°

$g = 16°$

0	296.9						
20	269.5	271.0	275.3	282.3	296.8	320.9	332.5
40	242.3	244.9	252.7	266.0	296.5	360.0	398.9
60	207.3	210.9	222.0	242.1	295.4	463.3	664.4
70	181.9	185.9	198.7	222.7	293.8	632.7	2298
80	144.7	149.0	163.1	191.5	289.2	1154	$1.1 \cdot 10^5$

$g = 18°$

0	374.4						
20	335.9	337.9	344.0	353.8	374.2	409.2	426.4
40	298.8	302.2	312.8	331.0	373.7	467.1	528.3
60	252.6	257.2	271.8	298.6	371.6	625.2	1007
70	220.3	225.4	241.8	273.2	368.6	879.3	7382
80	174.5	179.9	197.8	234.0	360.9	1535	$8.5 \cdot 10^4$

$\lambda = 0.50 \ \mu \ ; \ p = 0.85$

$g = 2°$

0	20.3						
20	20.0	20.1	20.1	20.2	20.3	20.5	20.6
40	19.7	19.8	19.9	20.0	20.3	20.7	20.9
60	19.2	19.3	19.4	19.8	20.3	21.2	21.7
70	18.7	18.8	19.0	19.4	20.3	21.9	22.7
80	17.7	17.8	18.1	18.7	20.3	24.1	26.9

$g = 4°$

0	32.7						
20	31.9	31.9	32.1	32.3	32.7	33.3	33.6
40	30.9	31.0	31.3	31.8	32.7	34.1	34.8
60	29.4	29.5	30.1	30.9	32.7	35.9	37.5
70	27.9	28.1	28.8	30.0	32.7	38.2	41.4
80	24.8	25.1	26.1	27.9	32.7	46.8	61.5

$g = 6°$

0	53.0						
20	51.1	51.2	51.5	52.0	53.0	54.5	55.1
40	48.8	49.0	49.7	50.8	53.0	56.6	58.3
60	45.3	45.7	46.8	48.7	53.0	61.2	65.8
70	42.1	42.6	44.1	46.7	53.0	67.5	77.7
80	36.0	36.6	38.6	42.4	53.0	93.4	173.6

$g = 8°$

0°	81.6						
20	77.7	77.9	78.5	79.6	81.6	84.7	86.1
40	73.3	73.7	75.0	77.1	81.6	89.2	93.0
60	66.6	67.3	69.6	73.1	81.6	99.4	110.5
70	60.9	61.8	64.5	69.3	81.5	114.2	143.1
80	50.6	51.7	55.1	61.7	81.3	178.8	775.5

$g = 10°$

0	119.0						
20	111.9	112.3	113.5	115.3	119.0	124.7	127.3
40	104.2	105.0	107.3	111.0	119.0	133.3	140.7
60	93.1	94.2	97.8	104.0	118.9	153.5	178.0
70	83.9	85.3	89.7	97.7	118.8	184.4	261.2
80	68.3	70.0	75.4	85.7	118.3	319.0	

Table 3 (cont.)

Zenith distance of point	Azimuth						
	0°	20°	40°	60°	90°	135°	180°
				$g = 12°$				
0	166.8						
20	155.1	155.8	157.7	160.7	166.8	176.6	181.2
40	142.7	143.9	147.5	153.5	166.8	191.7	205.3
60	125.4	127.2	132.7	142.4	166.6	228.7	279.8
70	111.8	113.9	120.5	132.7	166.2	287.6	490.1
80	89.9	92.3	100.1	115.3	165.1	528.3	$3.6 \cdot 10^5$
				$g = 14°$				
0	224.7						
20	206.4	207.4	210.3	215.0	224.7	240.3	247.7
40	187.7	189.5	194.9	204.0	224.5	265.1	288.4
60	162.6	165.1	173.1	187.4	224.0	328.2	430.1
70	143.7	146.7	155.9	173.4	223.3	431.5	979.5
80	114.6	117.9	128.5	149.6	220.9	805.8	$1.7 \cdot 10^5$
				$g = 16°$				
0	293.5						
20	266.5	267.9	272.2	279.2	293.5	317.3	328.8
40	239.6	242.1	249.9	263.0	293.2	355.9	394.3
60	204.9	208.4	219.4	239.3	292.1	458.1	656.1
70	179.7	183.7	196.3	220.1	290.5	625.9	2265
80	142.7	147.0	161.0	189.1	286.0	1145	$1.1 \cdot 10^5$

$$\lambda = 0.58 \ \mu; \ p = 0.89$$

Zenith distance of point	Azimuth						
	0°	20°	40°	60°	90°	135°	180°
				$g = 2°$				
0	15.1						
20	14.9	14.9	14.9	15.0	15.1	15.2	15.3
40	14.6	14.7	14.8	14.9	15.1	15.4	15.6
60	14.3	14.3	14.5	14.7	15.1	15.8	16.1
70	13.9	13.9	14.1	14.4	15.1	16.3	16.9
80	13.0	13.1	13.4	13.9	15.1	17.9	19.7
				$g = 4°$				
0°	27.4						
20	26.7	26.8	26.9	27.0	27.4	27.9	28.1
40	25.9	26.0	26.2	26.6	27.4	28.6	29.1
60	24.6	24.7	25.1	25.9	27.4	30.0	31.4
70	23.3	23.5	24.1	25.1	25.4	31.9	34.6
80	20.6	20.9	21.7	23.3	27.4	39.0	50.5
				$g = 6°$				
0	47.6						
20	45.8	45.9	46.2	46.7	47.6	48.9	49.5
40	43.8	44.0	44.6	45.6	47.6	50.8	52.3
60	40.6	41.0	42.0	43.7	47.6	54.9	58.9
70	37.7	38.2	39.5	41.9	47.5	60.5	69.5
80	32.0	32.6	34.5	37.9	47.5	83.7	152.0
				$g = 8°$				
0	76.1						
20	72.5	72.7	73.3	74.2	76.1	79.0	80.3
40	68.4	68.8	70.0	72.0	76.1	83.2	86.7
60	62.1	62.8	64.8	68.2	76.1	92.7	103.0
70	56.7	57.5	60.1	64.6	76.1	106.5	133.0
80	46.9	47.9	51.2	57.4	75.9	167.1	702.7

Table 3 (cont.)

Zenith distance of point	Azimuth						
	0°	20°	40°	60°	90°	135°	180°
$g = 10°$							
0	113.5						
20	106.8	170.1	108.3	110.0	113.5	119.0	121.4
40	99.4	100.1	102.3	105.8	113.5	127.1	134.2
60	88.7	89.8	93.2	99.2	113.4	146.3	169.5
70	79.9	81.2	85.5	93.1	113.3	175.8	247.9
80	64.8	66.4	71.6	81.5	112.8	305.8	
$g = 12°$							
0	162.1						
20	150.7	151.4	153.2	156.2	162.1	171.7	176.1
40	138.6	139.8	143.3	149.2	162.1	186.4	199.4
60	121.8	123.5	128.9	138.3	161.9	222.1	271.5
70	108.5	110.5	117.0	128.8	161.6	279.4	474.4
80	87.0	89.3	96.9	111.8	160.5	516.2	$3.6 \cdot 10^5$
$g = 14°$							
0	220.9						
20	202.9	203.9	206.8	211.4	220.8	236.3	243.5
40	184.4	186.2	191.6	200.6	220.7	260.5	283.4
60	159.7	162.3	170.1	184.2	220.3	322.6	422.2
70	141.1	144.0	153.2	170.3	219.5	424.2	959.0
80	112.3	115.5	126.0	146.9	217.1	795.8	$1.7 \cdot 10^5$
$g = 16°$							
0	291.8						
20	264.9	266.3	270.6	277.5	291.7	315.4	326.8
40	238.1	240.7	248.4	261.4	291.4	353.7	391.9
60	203.6	207.1	218.1	237.9	290.4	455.3	651.8
70	178.6	182.5	195.0	218.7	288.8	622.3	2247
80	141.7	145.9	159.8	187.8	284.3	1141	$1.1 \cdot 10^5$

$\lambda = 0.70 \ \mu ; \ p = 0.95$

	0°	20°	40°	60°	90°	135°	180°
$g = 2°$							
0°	9.9						
20	9.8	9.8	9.8	9.8	9.9	10.0	10.0
40	9.6	9.6	9.7	9.7	9.9	10 1	10.2
60	9.3	9.4	9.5	9.6	9.9	10.3	10.5
70	9.1	9.1	9.2	9.4	9.9	10.6	11.0
80	8.4	8.5	8.7	9.1	9.9	11.6	12.7
$g = 4°$							
0	21.8						
20	21.2	21.3	21.4	21.4	21.8	22.2	22.3
40	20.6	20.7	20.8	21.2	21.8	22.7	23.2
60	19.5	19.6	20.0	20.6	21.8	23.9	24.9
70	18.5	18.6	19.1	19.9	21.8	25.3	27.4
80	16.2	16.4	17.2	18.5	21.8	30.9	39.3
$g = 6°$							
0	41.5						
20	40.0	40.1	40.4	40.7	41.5	42.7	43.2
40	38.2	38.4	38.9	39.8	41.5	44.3	45.7
60	35.4	35.7	36.6	38.2	41.5	47.9	51.4
70	32.8	33.2	34.4	36.5	41.5	52.7	60.4
80	27.7	28.2	29.9	33.0	41.5	73.0	129.0

Table 3 (cont.)

Zenith distance of point	Azimuth						
	0°	20°	40°	60°	90°	135°	180°
$g = 8°$							
0	69.9						
20	66.5	66.7	67.3	68.1	69.9	72.5	73.7
40	62.7	63.1	64.2	66.0	69.9	76.4	79.6
60	56.9	57.5	59.4	62.6	69.9	85.0	94.3
70	51.9	52.7	55.1	59.3	69.8	97.6	121.4
80	42.7	43.6	46.7	52.5	69.7	153.7	622.4
$g = 10°$							
0	107.3						
20	100.9	101.3	102.3	104.0	107.3	112.5	114.8
40	94.0	94.6	96.7	100.1	107.3	120.2	126.8
60	83.8	84.8	88.1	93.7	107.2	138.3	160.0
70	75.3	76.6	80.7	87.9	107.1	166.0	232.9
80	60.9	62.4	67.3	76.8	106.7	290.7	
$g = 12°$							
0	156.0						
20	145.0	145.6	147.4	150.3	156.0	165.2	169.4
40	133.3	134.5	137.9	143.5	156.0	179.2	192.8
60	117.0	118.7	123.9	133.1	155.8	214.6	260.7
70	104.1	106.1	112.4	123.9	155.5	268.8	453.3
80	83.2	85.5	92.8	107.2	154.4	500.3	$3.6 \cdot 10^5$
$g = 14°$							
0°	215.6						
20	198.1	199.0	201.8	206.4	215.6	230.6	237.7
40	180.0	181.8	187.0	195.8	215.5	254.3	766.6
60	155.8	158.3	166.0	179.7	215.0	314.8	411.4
70	137.5	140.3	149.3	166.2	214.3	414.2	931.0
80	109.2	112.3	122.6	143.0	212.0	782.0	$1.7 \cdot 10^5$
$g = 16°$							
0	290.0						
20	262.3	263.8	268.0	274.8	288.9	312.4	323.7
40	235.8	238.3	245.9	258.9	288.6	350.3	388.1
60	201.6	205.1	215.9	235.5	287.6	450.9	644.9
70	176.7	180.6	193.0	216.5	286.0	616.6	2220
80	140.0	144.2	158.1	185.8	281.6	1133	$1.1 \cdot 10^5$

will have an inflection, after which h_0 will remain constant. This is due to the fact that solar rays that pass through the entire atmosphere at distances greater than 50 km are practically unattenuated.

The obtained data show that when defining the effective height h_0 of the twilight layer we cannot, as is commonly done, use a single value of h_0, since it depends upon the sun position g and the wavelength of the observed monochromatic radiation. Moreover, it obviously does not make sense to make observations in white light, since such observations could give a rather wide range of twilight-layer heights.

After determining h_0, we calculated the heights h of the twilight layer for the various points by formulas (2)–(6). These values for the four wavelengths are given in Table 3.

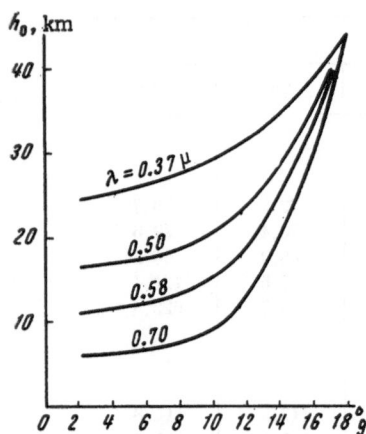

Fig. 2. Height h_0 versus sun
positioning.

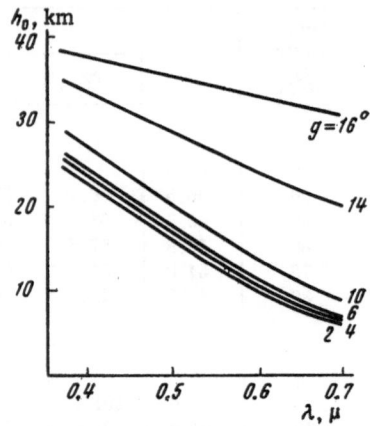

Fig. 3. Height h_0 versus wavelength.

As can be seen from this table, the heights of the twilight layer h differ greatly from point to point for the same sun position. Thus, at any given moment primary twilight at different points in the sky is determined by scattering in layers of different heights above the earth's surface. This accounts for the difference in the characteristics of radiation arriving from different points at the same moment in time.

Our results apply to a homogeneous model of a gaseous atmosphere with ozone. The presence of thick dust layers in the atmosphere can change the heights of the effective twilight rays. In this case, the "twilight ray scheme" may not be accurate enough. Integral equations must be solved in a rigorous examination, for example, of the problem of determining the dust concentration by measuring the brightness of the twilight sky.

LITERATURE CITED

1.   V. G. Fesenkov, Astron. zh., Vol. 7, p. 100 (1930).
2.   N. M. Shtaude, Trudy Komissii po izucheniyu stratosfery pri Akad. Nauk SSSR, Vol. 1, p. 1 (1936).
3.   N. M. Shtaude, Izv. Akad. Nauk SSSR, seriya geogr. i geofiz., Vol. 2, p. 349 (1947).
4.   N. B. Divari, Geomagnetizm i aeronomiya, Vol. 2, p. 720 (1962).
5.   N. B. Divari and L. I. Plotnikova, Astron. zh., Vol. 42, p. 1090 (1965).

EVALUATING THE DUST COMPONENT OF THE UPPER ATMOSPHERE BY THE TWILIGHT METHOD

A. E. Mikirov and A. A. L'vova

Today, because of the achievements in rocket engineering, the direct method has become the principal method for studying the atmosphere. Tables of standard atmospheres, summarizing the results of studies of structure and composition, give densities, pressures, temperatures, molecular weights, and other structural parameters up to very high altitudes for a great variety of geophysical conditions: time of day, season, solar cycle, etc. Other atmospheric parameters, including the optical properties of the upper atmosphere, have been studied just as thoroughly.

Before the advent of rockets, a number of indirect methods were used to study the properties and structure of the upper atmosphere. One such method was twilight sounding. This is a very old, traditional method for studying the structure and optical properties of the atmosphere at heights from about 20 to 120-150 km.

A mathematical theory of twilight as applied to single scattering was first advanced by V. G. Fesenkov [1]. The method has been improved and refined by many researchers. Later, V. G. Fesenkov studied the twilight phenomenon in greater detail. He took into account refraction of the sun's rays and the sphericity of the atmosphere [2, 3] and developed a way of allowing for secondary scattering [4, 5]. With the development of an experimental method for taking into account secondary scattering [5] and a number of other refinements [6], the twilight method has reached a certain degree of perfection. It is used today by many investigators both here and abroad [6, 24-31].

Let us consider the real possibilities of the twilight method in studying the optical properties of the upper atmosphere, particularly the structure of the earth's dust cloud. For this, we must examine, evaluate, and take into account all of its advantages and shortcomings, errors, and limitations.

The advantages of the twilight method are obvious. Since observations are made from the earth's surface, they are clearly less expensive than rocket studies, and in this case, it is simpler and easier to organize a network of observation stations for investigating the space and time properties of the dust cloud. It must be noted, too, that the procedure itself for twilight observations is not as simple as it was many years ago, since a number of additional observations are required for interpretation of the obtained data. Twilight observations, moreover, require a sky that is clear at all azimuths, and this happens rather rarely. For example, only five more or less clear days could be found out of a month of observations at the astronomical station Solnechnaya.

Measurements of twilight brightness yield decreasing curves that approach a limit — the brightness of the night sky — as the sun sinks below the horizon.

The measured brightness is the sum of many components.

1) Single scattering of direct solar radiation by an absolutely pure atmosphere.

2) Scattering of direct solar radiation by aerosols near the dust cloud.

3) Multiple scattering by an absolutely pure atmosphere and aerosols of the underlying layers (mainly tropospheric) illuminated by adjacent areas of the sky.

4) Natural luminescence, which can be divided into the luminescence from the sun and night glow.

In order to study the optical properties of the dust cloud, all components except dust scattering must be eliminated from the total brightness of twilight. Let us consider these components in greater detail.

The night glow can be determined experimentally if measurements are made up to the complete cessation of twilight and nightfall. Atmospheric luminescence from the sun must be eliminated beforehand, i.e., measurements must be made in a spectral region in which there are no atmospheric emissions.

Single scattering by air molecules can be eliminated rather reliably with the aid of additional calculations of Rayleigh scattering. Since the structure of the gaseous atmosphere is well-known today, only the annual and diurnal variations of density and the level of solar activity need be taken into account. According to many papers (for example, [7]), at 90–110 km there is practically no diurnal variation in atmospheric density, since the density change from day to night at these heights does not exceed 10%. The tables of standard atmospheres give the corresponding corrections for the annual variation and solar activity [8].

Multiple scattering is the most difficult and complicated problem of the twilight method. The accuracy with which multiple scattering is taken into account determines, in many respects, the accuracy of the method itself.

There are two approaches to the problem of multiple scattering in twilight — theoretical and experimental. Numerous papers by F. F. Yudalevich [9, 10], and papers by N. M. Shtaude, R. Robley, E. Halbert, and a number of others [11, 13] have been devoted to theoretical calculations.

Rigorous methods of solving the problem involve great mathematical difficulties. All calculations, therefore, are in the nature of estimates of secondary scattering as compared with primary.

A certain average structure of the atmosphere is presupposed, and the dependence of secondary scattering upon the position of the sun is mainly determined qualitatively.

Recently, J. V. Dave [14] calculated light scattering of up to seventh-order with a computer for the somewhat different problem of determining ozone concentration from the reversal effect. Calculations were made for solar zenith angles of 0 to 90°. Multiple scattering was found to be dependent upon the solar zenith angle and upon the atmospheric structure adopted. This method of taking into account multiple scattering probably has promise, although it is more difficult as applied to twilight, when the solar zenith angles are greater than 90°. And in calculations of this kind, one must know beforehand the photometric structure of the light source — the twilight segment — which is the ultimate aim of the twilight method.

V. G. Fesenkov is the author of an experimental method for determining multiple scattering in twilight. In this case, the twilight brightness is observed in the solar vertical circle at angles $\xi > 6°$ of the sun below the horizon at two symmetric points relative to the zenith 20° above the horizon. The total observed brightness at the point on the sun's side consists of the primary twilight segment (singly scattered light J_1, the tropospheric component), multiply scattered light J_2, and the more or less constant brightness of the night sky and zodiacal light J_3:

$$J = J_1 + J_2 + J_3.$$

At the other symmetric point, which is on the side opposite the sun, the observed brightness consists only of the tropospheric component J_2' and the brightness of the night sky and zodiacal light J_3', since for $\xi > 6°$ this point is in the earth's shadow and the direct rays of the sun do not reach it: $J' = J_2' + J_3'$. The values J_3 and J_3' are determined directly by experiment. If we find $J_2' = J' - J_3'$ from observations and determine the ratio $J_2/J_2' = k$ theoretically (or experimentally), we can find the brightness of the primary twilight segment, which is directly related to the optical properties of the upper atmosphere:

$$I = J - J_3 - k(J' - J_3'), \tag{1}$$

where k is the ratio of the effects of multiply scattered light at the two symmetric points and is a function of the scattering indicatrix. In his theoretical calculations of k, V. G. Fesenkov used the standard scattering indicatrix, as measured under daytime conditions, the shape of which remains unchanged over a wide range of variation of the optical characteristics of the troposphere.

Tropospheric scattering in twilight is a reasonable approximation of daytime illumination of the clear sky. The twilight ray when $\xi > 6°$ is the external light source with respect to the scattering atmosphere as the sun is in the daytime. Only one source (the sun) can be considered a point source; the other source (the twilight segment) is an extended source and requires integration over its entire area. The desired factor k can be represented theoretically by the expression

$$k = \frac{\iint B(\xi, A) f(\vartheta_1, z_1, \xi) \sin \xi \, d\xi \, dA}{\iint B(\xi, A) f(\vartheta_2, z_2, \xi) \sin \xi \, d\xi \, dA}, \tag{2}$$

where

$$f(\vartheta) = 1 + 5.5 (e^{-3\vartheta} - 0.009) + 0.5 \cos^2 \vartheta$$

is the tropospheric indicatrix [15] and $B(\xi, A)$ is the brightness of the primary twilight segment at some point.

As can be seen from this formula, the brightness of the primary twilight segment is not of crucial importance, since it appears in the numerator and denominator in the same way. V. G. Fesenkov calculated it under the simplifying assumptions that refraction and Rayleigh scattering, which is proportional to the gaseous density of the atmosphere at a given height, were absent.

Below are values of the factor k for a number of solar zenith angles with and without allowance for ozone absorption:

ξ	96	98	100	102	104	106°
k without ozone	2.02	1.94	1.86	1.78	1.77	1.77
k with ozone	2.13	2.13	2.13	2.12	2.13	2.14

The existence of a criterion for estimating the weight and confidence of each point on the twilight curve must be a necessary condition for the objectivity and, therefore, the applicability of the twilight method. This is necessary in order that random decreases in brightness due to small clouds or any other shielding obstacles anywhere in the area of the terminator will not simulate attenuating layers.

G. V. Rozenberg [16] suggests that this be done by additional observations of atmospheric transmittance for the entire twilight period. Then, according to the shape of the transmittance curve, the points that do not fit into the general shape of the twilight curve without a corresponding change in transmittance can be eliminated.

With an objective criterion for estimating the confidence of points on the twilight curve, the accuracy of the twilight method is determined by 1) the accuracy with which multiple scattering is taken into account and 2) the accuracy with the data are tied in to the height of the effective scattering layer according to the depth of the sun below the horizon.

Let us examine these two conditions separately.

I. According to Fesenkov's method, the brightness of the primary twilight segment is determined by formula (1)

$$I = J - J_3 - k(J' - J_3').$$

Under nighttime conditions, usually $J_3 = J_3'$. But if $J_3 \neq J_3'$, as, for example, in our measurements (see below), they still differ little. Since J_3 and J_3' are constant and their effect begins to be felt only in very deep twilight, they can be ignored when estimating the accuracy with which secondary scattering is taken into account. Then

$$I \simeq J - kJ'.$$

The factor k depends little upon the sun's position (see above); on the average, $k = 2$.

Now the relative error in determining the brightness of the primary twilight segment can be written as

$$\frac{\Delta I}{I} = \frac{\Delta J + k\Delta J'}{J - kJ'} = \frac{\dfrac{\Delta J}{J} + k\dfrac{\Delta J'}{J}}{1 - k\dfrac{J'}{J}}, \tag{3}$$

where ΔJ and $\Delta J'$ are the absolute errors in brightness at the two symmetric points on the solar vertical circle.

We shall express the relative errors in determining brightness at the symmetric points by means of the known instrumental error in measuring brightness δ', which introduces a certain averaging into their values. Then

$$\frac{\Delta I}{I} = \delta = \frac{\delta' + k\delta'\left(\dfrac{J'}{J}\right)}{1 - k\dfrac{J'}{J}}. \tag{4}$$

The error of the method is a function of the ratio of the brightnesses at the symmetric points. The error increases greatly as J approaches J', and when $kJ' = J$, the error becomes infinite.

To increase the accuracy of determining the brightness of the primary twilight segment, we must reduce the instrumental error, and also operate in the area of $J'/J < 1/k$, which imposes limitations on the heights and solar zenith angles that can be measured. Large angles are *a priori* eliminated, while angles $\xi < 6°$ are eliminated by the measurement method itself.

The derivative of the intensity dI/dh is required to determine the dust concentration or scattering factor at a given height. In practice, this derivative is found by means of finite differences, i.e.,

$$\frac{dI}{dh} = \frac{\Delta I}{\Delta h} = \frac{I_{h_1} - I_{h_2}}{h_1 - h_2}. \tag{5}$$

This increases the error of the twilight method. The error of expression (5) has the form:

$$\delta_h = \frac{2\frac{\Delta I}{I}}{1 - I_{h_1}/I_{h_2}}, \tag{6}$$

where $\Delta I/I$ is determined by formula (4).

After the derivative has been found, the total error of the twilight method has more than doubled.

II. The main parameter that determines the tying in of the twilight data to height is the perigee of direct solar radiation for a given wavelength λ. The perigee of solar rays of a given wavelength λ is the point of inflection of the atmospheric-transmittance curve $T(\lambda, h)$ [6]. The transmittance $T(\lambda, h)$ or optical thickness $\tau(\lambda, h)$ near the perigee is determined by the optical structure of this region. The perigee of solar rays during twilight is located in the troposphere and the lower layers of the stratosphere, which are greatly affected by weather. A rigorous formulation of the twilight problem requires, therefore, that the form of the transmittance or optical-thickness function over the entire height range be determined experimentally every time twilight brightness is measured. These data should be obtained independently, for example, from measurements of the brightness of artificial satellites as they emerge from the earth's shadow, of the brightness of the sun at twilight at various heights from rockets or balloons, etc.

So far, to determine the perigee of solar rays h_0, most researchers have used a "standard" function $T(\lambda, h)$, theoretically calculated on the basis of models of the atmosphere. This is hardly justified, especially because, as a rule, aerosol extinction is ignored in such calculations.

G. V. Rozenberg [6], proceeding from the specific nature of the function $T(\lambda, h)$, which preserves its form with wide variation of the optical state of the atmosphere, proposed a method for determining the optical thickness of the atmosphere $\tau(\lambda, h)$ from the brightness of the twilight sky at certain points (the method of effective shadow heights).

According to this method,

$$\tau(\lambda, h) = \frac{\ln m + \ln S_i(z = 0, \xi, \lambda) - \ln S_i\left(z, A = \pm\frac{\pi}{2}, \xi, \lambda\right)}{m - 1}, \tag{7}$$

where m is the atmospheric mass and S_i is the brightness at the zenith ($z = 0$) and at points on the vertical circle perpendicular to the solar vertical circle ($A = \pm\pi/2$).

The error in finding height h_0 by the method of effective shadow heights is determined by the error in the optical thickness $\tau(\lambda, h)$ and by its subsequent double differentiation with respect to height, which increases this error considerably.

It follows from formula (7) that the error in the optical thickness $\tau(\lambda, h)$ is determined by the error in measuring twilight brightness at the corresponding points as well as by the relationship of these brightness values themselves. If we know S_i at $z = 0$ and $A = \pm\pi/2$, we can find the error in $\tau(\lambda, h)$ and then the error in h_0.

We did not make twilight measurements by the method of effective shadow heights, so in estimating the corresponding error we shall cite [6].

According to [6], the perigee of solar rays h_0 is determined with an accuracy of ± 3 km, which means that the error in the twilight brightness after they have been tied in to height could be greater than 50%. This error characterizes the degree of accuracy of the twilight method.*

The error in determining the dust concentration or scattering factor at a given height can be calculated by formula (5).

As an example, let us consider the results of twilight measurements that we made in August 1963.

We estimated the error due to tying in the twilight data to height according to [6].

The brightness of the twilight sky was measured by V. G. Fesenkov's method at the astronomical station Solnechnaya (2200 m above sea level) near Kislovodsk. The measurements were made with an FIR-3 rocket photometer, which is described in [16]. The sensitivity threshold of the instrument is $2 \cdot 10^{-10}$ W/cm$^2 \cdot$ sr $\cdot \mu$ and its dynamic range is over three orders of magnitude. Neutral filters provide an additional range, by attenuation, of four orders of magnitude. The viewing angle is two degrees.

The measurements were made with an interference filter with a wavelength of 5200 Å and a spectral half-width $\Delta\lambda = 100$ Å.

In the spectral region of 5200 Å, the natural emission of the atmospheric line of atomic nitrogen (transition $^2D \rightarrow {}^4S$), which has a wavelength of 5199 Å, must be taken into account.

This line was measured under twilight conditions in 1958 by Blackwell et al. [18] and somewhat earlier by Dufay [19]. In [18], it is assumed that the maximum intensity of luminescence in the 5199 Å line during twilight does not exceed 600 Rayleighs. If we take this maximum value of 600 Rayleighs as the initial value, then at 5200 Å when the filter has a spectral half-width $\Delta\lambda = 100$ Å and in the absence of other light sources, we obtain an energy of $1.5 \cdot 10^{-11}$ W/cm$^2 \cdot$ sr $\cdot \mu$.

The minimum energies that we measured during twilight exceed by at least two orders of magnitude the energy of the 5199 Å line adjusted to the spectral interval $\Delta\lambda = 100$ Å. Therefore, the twilight energies that we measured were mainly a result of scattering rather than of the luminescence of the atmosphere.

Now let us consider the experimental data. As an example, Fig. 1 shows experimental curves for both symmetric points on the solar vertical circle for the morning of 20 August 1963. The solar zenith angle is plotted on the axis of the abscissas and the brightness is plotted on the axis of the ordinates. The curves for the other days are of the same nature. As can be seen from the figure, the two curves are not always parallel and the experimental points have a slight spread. With nightfall, both curves run parallel to the axis of the abscissas and do not coincide with one another. This is probably due to the presence of the zodiacal cone in one direction and is a reflection of the actual heterogeneity in the brightness of the night sky.

----

*V. K. Pyldmaa [32] has used Rozenberg's method [6] to study twilight phenomena. It is shown in [32] that the method for determining transmittance proposed in [6] cannot be used with solar zenith angles of from 96 to 100°, i.e., during the principal period of twilight observation. Other methods for determining the perigee of solar rays h_0 can scarcely give accuracy better than ± 3 km. Therefore, the error in determining twilight brightnesses due to tying them to height will be 50%, as a rule.

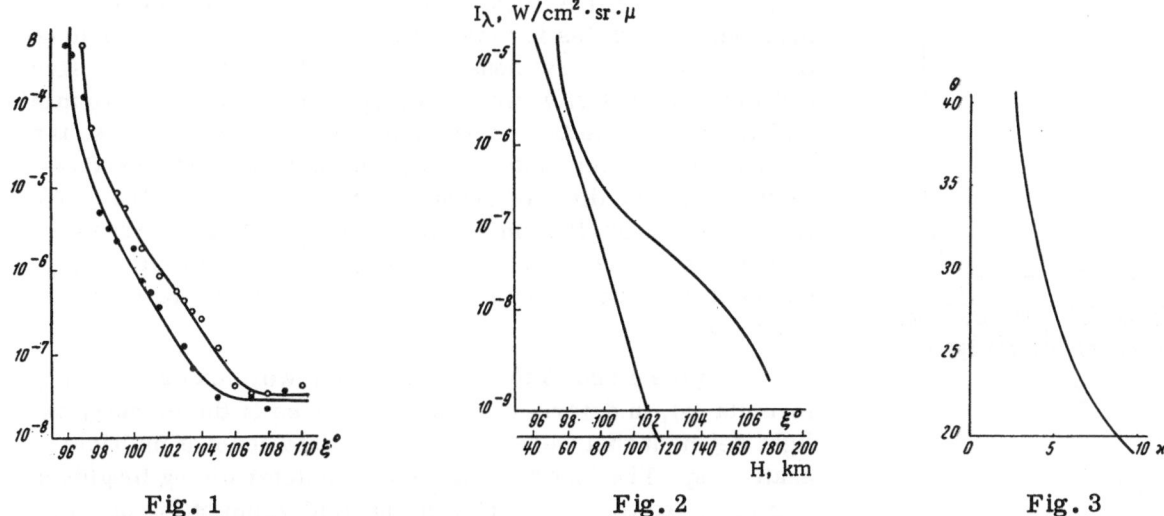

Fig. 1   Fig. 2   Fig. 3

Figure 2 shows the intensity of the primary twilight segment in $W/cm^2 \cdot sr \cdot \mu$ for the same day as calculated by formula (1) without allowance for the effect of ozone on the factor k. The height above the earth's surface together with the solar zenith angle is plotted on the axis of the abscissas in this figure. A reference curve (left) for Rayleigh scattering is also given.

The intensity of the primary twilight segment can be written as

$$I_\lambda = I'_\lambda + I''_\lambda,$$

where

$$I'_\lambda = \frac{3}{16\pi} S_{0\lambda} (1 + \cos^2 \varphi) m\tau_p \tag{8}$$

is for molecular scattering and

$$I''_\lambda = \frac{S_{0\lambda}}{\pi} \eta(\rho) \overline{\varkappa}(\varphi, \rho) m\tau_a \tag{9}$$

is for aerosol scattering.

In formulas (8) and (9), τ_p and τ_a are the molecular and aerosol optical thicknesses; $S_{0\lambda}$ is the luminous flux from the sun in the measured spectral interval [20]; and $\overline{\varkappa}(\varphi, \rho)$ is the mean scattering indicatrix for dust, which was calculated by Smerkalov [21, 22] with the Mie formulas for $\rho = 6$.

If it is assumed that the dust does not absorb light and that the size of the dust particles is such that they scatter light of 5200 Å best, then the lower limit of the dust concentration at any given height can be determined. The size of the scattering particles is determined in this case, since for $\lambda = 5200$ Å, n = 1.5 (silicon), and $k(\rho) = 4$, $\rho = 2\pi r/\lambda = 6$, and r = 0.5 μ. The dust concentration $N_n(h)$ can be determined from

$$\frac{dI''_\lambda}{dh} = \frac{d}{dh}(I_\lambda - I'_\lambda) = \frac{S_{0\lambda}}{\pi} \eta(\rho) \overline{\varkappa}(\varphi, \rho) mk(\rho) \pi r^2 N_n(h). \tag{10}$$

The change in the angle of the scattering indicatrix (for Rayleigh and aerosol scattering) as the sun sank below the horizon was taken into account in the calculations. As the solar zenith angle varies from 96 to 110°, the scattering angle varies from 26 to 40°. Figure 3 shows the aerosol scattering indicatrix versus the scattering angle for $\rho = 6$.

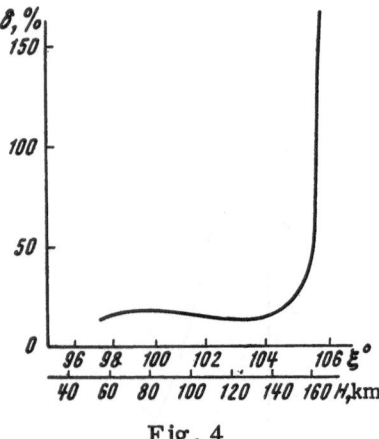

Fig. 4

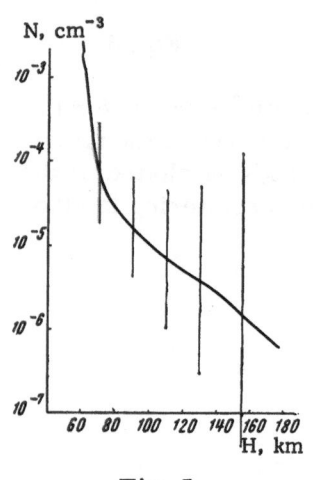

Fig. 5

In calculating Rayleigh scattering, we used the standard atmosphere of CIRA 1965, which gives the annual mean atmospheric densities for mean solar activity F = 150 for heights of 40-120 km. Corrections were made for the seasonal effect and for solar activity, which in August 1963 was close to minimal (F = 78). The two corrections had opposite signs, so that after they had been made, the density used to calculate Rayleigh scattering differed from the standard by less than 10% for heights of 40-100 km and by 20-30% for heights greater than 120 km, where the density and, therefore, the Rayleigh component become very small. The solar data were taken from [20].

Figure 4 shows the errors, which were calculated by formula (4), in determining the brightness of the primary twilight segment that were due to taking into account multiple scattering. The instrumental error in determining brightness was assumed to be 5%. This is the real value of the error, since at the present time it is difficult to calibrate the photoelectric instrument in absolute units with high accuracy. The error curve approaches infinity when the solar zenith angle is about 105° (150 km). Therefore, we can determine the brightness of the primary twilight segment only to heights less than 150 km. At zenith angles of $96° < \xi < 104°$, the error in determining the brightness of the primary twilight segment is about 20% in the best case; on some days, it reaches 50%.

As was pointed out above, the error in brightness due to tying the twilight data to height is 50%.

The errors in determining concentration are very small and depend upon the size of the step of differentiation with respect to height. The errors increase as the step is reduced.

When only multiple scattering is taken into account, the errors in dust concentration reach fourfold, and when only tying to height is taken into account, they reach sixfold. When both components are taken into account, the total error reaches tenfold.

Figure 5 shows the results of calculations of dust concentration by formula (10) as a function of height. It can be seen that in general the dust concentration decreases with height.

The total error in determining the concentrations is also given in Fig. 5. This error is so great that at a given height the dust concentration can be indicated only with accuracy of plus or minus one order of magnitude.

We thank all those who assisted in this work: N. M. Tikhonova, A. E. Khaletskii, O. K. Kudryavov, and N. P. Bobkov.

LITERATURE CITED

1. V. G. Fesenkov, Trudy Glavnoi Rossiiskoi astrofiz. obs., Vol. 2 (1923).
2. V. G. Fesenkov, Doklady Akad. Nauk SSSR, Vol. 101, No. 5 (1955).
3. V. G. Fesenkov, Astrofiz. zh., Vol. 36, No. 2 (1952).
4. V. G. Fesenkov, Astron. zh., Vol. 32, p. 265 (1955).
5. V. G. Fesenkov, Trudy Astrofiz. inst. Akad. Nauk Kazakh SSR, Vol. 3, p. 214 (1962).

6.   G. V. Rozenberg, Twilight [in Russian], Fizmatgiz (1963) [English translation, Plenum Press, New York (1966)].

7.   A. C. Faire and K. S. W. Champion, Space Res., Vol. 5, p. 1039 (1965).

8.   CIRA (1965).

9.   F. F. Yudalevich, Doklady Akad. Nauk SSSR, Vol. 75, p. 6 (1950).

10. F. F. Yudalevich, Izv. Akad. Nauk SSSR, seriya geogr. i geofiz., Vol. 6 (1950).

11. N. M. Shtaude, Doklady Akad. Nauk SSSR, Vol. 59, No. 7 (1948).

12. R. Robley, Ann. Geophys., Vol. 80, p. 1 (1952).

13. E. Hulbert, J. Opt. Soc. America, Vol. 28, No. 7 (1938).

14. J. V. Dave, J. Atmos. Sci., Vol. 22, No. 3 (1965).

15. E. V. Pyaskovskaya-Fesenkova, Investigation of Light Scattering in the Earth's Atmosphere, Izd. Akad. Nauk SSSR (1957).

16. A. E. Mikirov, Iskusstvennye sputniki Zemli, No. 13 (1962).

17. A. E. Mikirov, Kosmicheskie issledovaniya, Vol. 3, No. 2 (1965).

18. D. E. Blackwell, M. F. Inghain, and H. N. Rundle, Ann. Geophys., Vol. 16, No. 1 (1960).

19. Dufay, Ann. Phys., Vol. 8, p. 813 (1953).

20. F. S. Johnson, J. Meteorol., Vol. 11, p. 431 (1954).

21. V. A. Smerkalov, Trudy VVIA im. Zhukovskogo, No. 871 (1961).

22. V. A. Smerkalov, Trudy VVIA im. Zhukovskogo, No. 986 (1962).

23. L. G. Jacchia, Space Res., Vol. 5, p. 1152 (1965).

24. N. M. Shtaude, Trudy Komissii po izucheniyu stratosfery pri Akad. Nauk SSSR, Vol. 1 (1936).

25. N. M. Shtaude, Izv. Akad. Nauk SSSR, seriya geogr. i geofiz., Vol. 13, No. 4 (1949).

26. T. G. Megrelishvili, Izv. Akad. Nauk SSSR, seriya geogr., No. 3 (1956).

27. T. G. Megrelishvili, Izv. Akad. Nauk SSSR, seriya geogr., No. 4 (1958).

28. N. B. Divari, Geomagnetizm i aeronomiya, Vol. 4, p. 886 (1964).

29. N. B. Divari and L. I. Plotnikova, Astron. zh., Vol. 42, No. 5 (1965).

30. F. E. Volz and R. M. Goody, J. Atm. Sci., Vol. 19, p. 385 (1962).

31. V. K. Pyldmaa, Izv. Akad. Nauk SSSR, seriya fizika atmosfery i okeana, Vol. 1, No. 11 (1965).

BRIGHTNESS AND POLARIZATION OF TWILIGHT AT THE CELESTIAL POLE

D. I. Stepanov and L. B. Gusakovskaya

The method of determining the tropospheric component from observations at two symmetric points on the solar vertical circle [1] cannot be used for the early period of twilight, when $\zeta_\odot > 6\text{-}7°$, when the primary twilight segment is not yet detached from the horizon on the antisolar vertical circle or is at a negligible distance from it [2]. If it is assumed that during bright twilight the degree of higher-order polarization is negligible or entirely absent [3], the tropospheric component can be determined from observations of the brightness and polarization of the twilight sky at points not on the solar vertical circle far from the zenith. With this in mind, we chose for observations the region near the celestial pole. At the latitude of Dushanbe, this region of the sky is fairly far removed from the solar vertical circle and from the zenith. Therefore, the observations were made in a region even closer to the horizon, but besides taking into account the effect of the tropospheric component for the period of early twilight, we studied the variation of the degree of polarization, brightness, and angle of polarization at the celestial pole.

Observations at the celestial pole were made concurrently with observations of chromatic polarization on the solar vertical circle during an expedition in Tadzhikistan with a twilight photoelectric photometer, which is described briefly in [4]. The brightness as obtained through yellow and ultraviolet filters and the polarization were recorded successively on the same channel of the photometer.

Figure 1 shows the degree of polarization of twilight as a function of the dip angle of the sun below the horizon at the celestial pole for several days in 1965.

As can be seen from the graph, the formation of twilight on those days up to $h_\odot \approx 10°$ was basically the same, since all curves display the same regularities. During early twilight, when $h_\odot \approx 5\text{-}6°$, with rare exceptions the polarization takes values within 0.45–0.55. A single or double minimum at $h_\odot \approx 8\text{-}10°.5$ is clearly seen on all curves. In most cases, polarization drops when the sun is more than $10°.5$ below the horizon. An abrupt decrease in polarization occurred at the end of July and the beginning of August at large solar zenith angles (curves 1, 2, 3, and 6). An interesting peculiarity that is unusual for zenith observations was a considerably lower polarization for the entire period of morning twilight as compared with evening. The polarization curve for morning necessarily had the same shape as the evening curve, but it was shifted somewhat to the low side and much lower overall than the evening curve. The relationships between the polarizations for morning and evening vary from day to day. The greatest difference in P for morning and evening was observed at the end of July and the beginning of August, while in June it was much smaller. Analysis of a number of curves showed that the

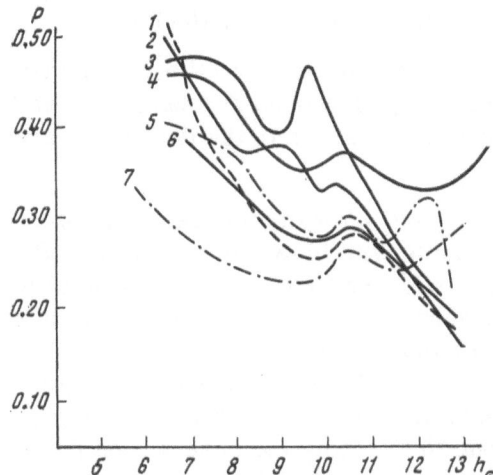

Fig. 1. Degree of polarization versus dip angle of sun below horizon (evening): 1) 23-24 July; 2) 6-7 August; 3) 5-6 August; 4) 29-30 July; 5) 26-27 July; 6) 29-30 June; 7) 13-14 July.

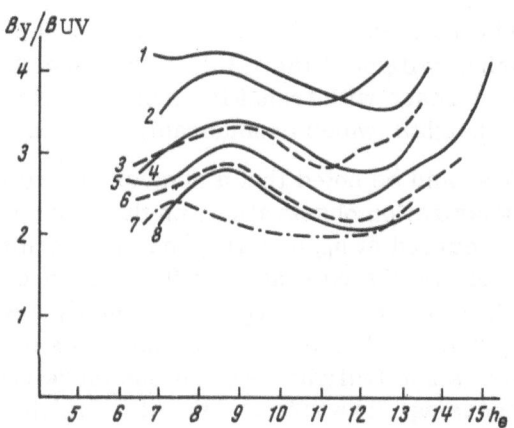

Fig. 2. Ratio of twilight brightnesses for yellow and ultraviolet filters versus dip angle of sun below horizon: 1) 4-5 August; 2) 6-7 August; 3) 26-27 July; 4) 29-30 July; 5) 7-8 July; 6) 26-27 June; 7) 29-30 June; 8) 23-24 June.

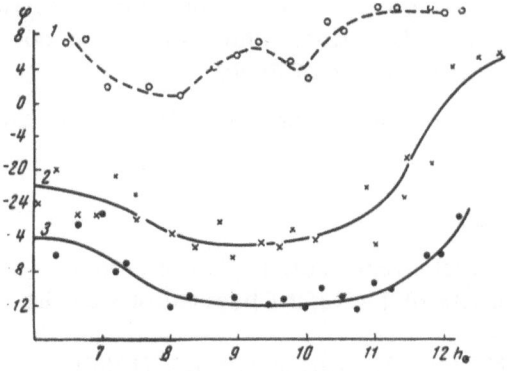

Fig. 3. Position of polarization versus dip angle of sun below horizon: 1) 27 June (evening); 2) 29-30 June (morning); 3) 4-5 July.

above-mentioned shift occurred in the evening polarization (this polarization increased gradually toward August). There was no such change in morning polarization from June to August. In August when $h_\odot \approx 8°$, polarization dropped greatly, which was easily identified with the growing meteor activity and the diurnal variation of the number of meteors that had been falling out (the peak was during the morning hours).

Simultaneous recording on the polarization channel of the brightnesses in the yellow B_y and ultraviolet B_{uv} regions made it possible to analyze the variation of the ratio on the same days (Fig. 2). As is apparent, the ratio, just as the polarization, increased from June to August, i.e., the brightness either increased in the yellow region or decreased in the ultraviolet.

The observed clearly expressed yellowing of twilight toward August could have been due to the annual variation of meteor activity. Additional dust particles distributed over a wide range of heights could possibly also have caused such an increase in the ratio B_y/B_{uv}. As was indicated above, however, for evening twilight on the same days from June to August up to $h_\odot \approx 10-11°$ there was an increase in the degree of polarization, which cannot be explained in terms of the effect of minute dust particles. In the range $h_\odot \approx 7-10°$, the ratio B_y/B_{uv} increased on all curves, while on the polarization curves minimum polarization was observed in this range of angles. Similar relationships were observed for later twilight.

The yellowing and the great decrease in the degree of polarization were evidently due to minute aerosol particles, which many researchers assume exist above 100 km.

The increase in the ratio B_y/B_{uv} for $h_\odot \approx 7\text{-}10°$ was possibly due to the increasing intensity during twilight of the red line 5300 and 6364 Å and the yellow doublet 5893 Å, whose brightnesses decrease with transition from twilight to night [5], as well as to the dust layer present at these heights, which considerably reduces polarization.

It should be noted that a number of regularities showed up on the curves of B_b/B_r versus $h_\odot$ in polarization observations on the solar vertical circle with blue and red filters. Reddening occurred at $h_\odot \approx 8\text{-}10°$, and the morning curves were shifted in the direction of lower $h_\odot$, but no clearly expressed shifting of the curves toward August was observed. In the antisolar direction the morning sky is usually always bluer than in the solar direction, but in the evening at the end of June the opposite was observed several times. The reason for this is still not clear, since twilight observations made under good atmospheric conditions were selected for processing. The effect of the city, which was about 20 km from the observation point and not on the solar vertical circle during twilight, must have manifested itself identically on all curves with close dates.

It was interesting to examine the variation in the angle of polarization at a point close to the celestial pole. For this, a relay mounted on the continuously rotating polaroid recorded on a tape the position of the polaroid every 60°. The degree of polarization and the angle of polarization were determined by V. G. Fesenkov's method from the brightnesses of the sky at three positions of the polaroid 60° apart [6]. Figure 3 shows a few curves of the angle of polarization as a function of the dip angle of the sun below the horizon. At first, the angle φ follows the sun, and then it begins to change more slowly. The slight increases or decreases of φ within 2-4° could be due to measurement error. The angle φ becomes constant in the range $h_\odot \approx 7.5\text{-}11°$, after which it begins to change in the opposite direction. The angle φ was observed to change in the range $h_\odot \approx 7\text{-}11°$ in some cases (evening of 27 June).

The angle φ was not determined for $h_\odot > 12\text{-}13°$, since the brightness of the sky became negligible and the error would have been too great.

LITERATURE CITED

1. V. G. Fesenkov, Izv. Astrofiz. inst. Akad. Nauk Kazakh SSR, Vol. 12, No. 3 (1961).
2. D. I. Stepanov and L. B. Gusakovskaya, Some Results of Twilight Observations on the Solar Vertical Circle [in Russian], Moscow (1967).
3. E. V. Pyaskovskaya-Fesenkova, Doklady Akad. Nauk SSSR, Vol. 123, No. 2 (1958).
4. D. I. Stepanov and L. B. Gusakovskaya, in: Meteor Radio Wave Propagation [in Russian], No. 2, Kazan' (1964).
5. N. B. Divari, Izv. Akad. Nauk SSSR, seriya geogr. i geofiz., Vol. 13 (1949).
6. V. G. Fesenkov, Astron. zh., Vol. 12, No. 4, p. 309 (1935).

PHOTOELECTRIC PHOTOMETRIC STUDY OF TWILIGHT IN THE INFRARED REGION

E. V. Gnilovskoi

Twilight, i.e., the aggregate of optical effects in the atmosphere in the transition period between day and night, is among the as yet imperfectly understood phenomena of nature. The scattering of sunlight in the atmosphere underlies the physics of twilight.

The foundations of the theory of twilight were laid by V. G. Fesenkov in the 1920s [1, 2]. After this, a great deal of observation time was devoted to twilight sounding of the atmosphere by the photoelectric method. Researchers usually limited themselves to studying the variation in the brightness of the twilight sky at the zenith, as a function of the dip angle of the sun below the horizon in relative units [3-6]. Later, such observations were made in particular spectral regions using filters [7-9]. The problem of the interference introduced by secondary scattering of light in the atmosphere remained unclear in the theory of twilight sounding.

In order to assess the role of secondary scattering and determine the effectiveness of the twilight method, I. A. Khvostikov and T. G. Megrelishvili [10, 11] compared the data obtained by twilight sounding with the data obtained by other methods. They found that the results of twilight sounding were in agreement with the results of the other methods, particularly rocket sounding, to heights on the order of 90 km. The results are in agreement to heights on the order of 120 km when corrections for secondary scattering are made.

Another aspect of twilight research is the color variation of the twilight sky as the position of the sun relative to the horizon changes. Numerous studies [7, 10-12], which were made mainly in the visible region of the spectrum, allow it to be considered an established fact that when the sun is below the horizon, starting with 3–4°, "bluing," i.e., a relative intensification of the shortwave components in the twilight sky, is observed. When the sun is 9-11° below the horizon, the blue is replaced by reddening. These effects have been studied by Megrelishvili [10], Megrelishvili and Khvostikov [11], Volz and Goody [13], and others using filters in the vislbe region. The color variation of the twilight region in the $1-\mu$ region has been studied by E. D. Sholokhova and M. S. Frish [23].

There is a great deal of interest in analyzing regularities in the absorption bands of atmospheric gases in the spectrum of the twilight sky.

The ozone absorption bands have been studied by N. B. Divari, and also by Volz and Goody. Volz and Goody have indicated that the total ozone content can be determined from the absorption bands in the spectrum of the twilight sky.

As far as we know, the oxygen and water-vapor absorption bands in the infrared region of the twilight spectrum have not been studied at all.

119

Most of the above-mentioned work was performed with filters, i.e., in specific regions of the spectrum. Such measurements do not provide a complete representation of the spectrum of the twilight sky.

A number of researchers have obtained spectra of the twilight sky by means of photographic photometry, usually with long exposures, which is equivalent to averaging over a wide range of solar zenith distances. Therefore, these measurements cannot be used to analyze the spectrum of the twilight sky and its time variation.

Extremely few spectra have been obtained by photoelectric photometry. In 1960, such spectra in the visible region were obtained at the Department of Atmospheric Optics of the Institute of Atmospheric Physics of the Academy of Sciences of the USSR under the direction of G. V. Rozenberg [19]. But as far as we know, the literature contains no data on the infrared spectrum of the twilight sky.

The results of experimental observations provided a basis for the further development of the theory of twilight by G. V. Rozenberg [19]. An important aspect is his introduction of the notion of the effective shadow boundary. G. V. Rozenberg's theoretical concepts provide qualitative explanations of principal twilight phenomena such as purple light, the belt of Venus, glow, etc.

But the theory of twilight is far from complete. The study of twilight should be considered primarily a method that enables us to find out about the structure and properties of the earth's atmosphere.

Future photometric observations of twilight will help to explain the laws of light scattering in the atmosphere and to answer questions about the aerosol composition of the atmosphere, the contribution of higher-order scattering to the overall brightness of the sky, regularities in the absorption bands of atmospheric gases, the height distribution of atmospheric gases, etc.

The aim of the work described here was to obtain and study the twilight-sky spectra at the zenith and at z = 60° in the near infrared region. A wavelength-scanning spectrophotometer with photoelectric recording in absolute units was used. Photoelectric photometry, with its inherent fast recording of luminous flux is more effective than photographic photometry for analyzing such a transient phenomenon as twilight. Moreover, the photoelectric detector has the advantage of being rather simple to calibrate in absolute units.

Our observations in 1963 were made in the range of sun heights of from +6 to −6°. The upper limit was chosen so that the beginning of the process could be caught, since twilight begins long before the sun goes below the horizon (see [19]). The lower limit was determined by the sensitivity of the apparatus.

The infrared region of the twilight spectrum was chosen for study for the following reasons.

1) This region had been studied little previously; in the literature known to us, there are no data on twilight spectra in the region of $0.6-1\,\mu$.

2) There is more of a basis for ignoring higher-order scattering in the near infrared region than in the visible, i.e., the effects caused by single scattering can be traced in purer form.

3) The establishment of regularities in the oxygen and water-vapor absorption bands in this region is of considerable interest.

4) Atmospheric transmittance is greater in the infrared than in the visible region.

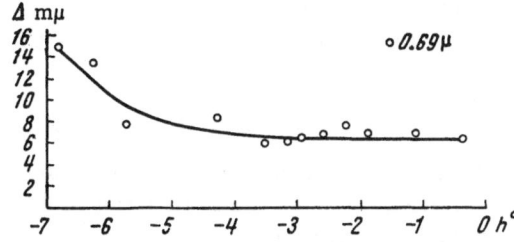

Fig. 1. Half-width of oxygen absorption band (0.69 μ) versus height of sun. Crimean Astrophysical Observatory, 24 September 1963.

5) The color effects of twilight in the infrared region is of particular interest.

Twilight spectra in the region of 0.6–1.1 μ in absolute units were obtained for the first time during the winter and spring of 1963 at the Pulkovo Astronomical Observatory [20]. We made more extensive observations in September and October of 1963 at the Crimean Astrophysical Observatory of the Academy of Sciences of the USSR. In all, about 300 twilight spectra were obtained.

The spectral system employed a monochromator with a Fastie diffraction grating. The monochromator had a luminosity of 1/9 and the lined part of the grating was 70 × 80 mm. The grating had 600 lines per 1 mm. The spectra were scanned by rotating the grating. In the infrared region (0.6–1.1 μ) the grating operated in the first order. KS-1 red glass was placed in front of the slit to cut off the second-order spectrum. The dispersion of the monochromator was 22 Å/mm. The detector was a combination of an FKT-1 image converter with a cesium oxide photocathode and a photomultiplier with an antimony-cesium cathode.* The image-converter cathode was cooled with dry ice. After amplification, the output signal of the photomultiplier went to a recorder. Since the kick of the recorder pen was proportional to the luminous flux incident on the detector and the dispersion of the monochromator with the diffraction grating depended little upon wavelength, the brightness of the twilight sky as a function of wavelength was obtained directly on the rectangular coordinate axes of the graph paper. The deviation from linearity of the receiving system did not exceed 1%. The apparatus was calibrated in absolute units with a standard lamp. The accuracy of measuring twilight-sky brightness was on the order of 20%. The sensitivity was changed by switching amplification ranges and varying within narrow limits the working voltage of the image converter.

The observation results showed that the presence of deep oxygen and water-vapor absorption bands whose rotational structure could not be resolved was most characteristic of the obtained spectra. The bands could not be resolved because of the low resolving power of the receiving apparatus, and also possibly because the rays under twilight conditions penetrate a great thickness of absorbing substance, which spreads the bands.

Let us discuss some of the regularities in the oxygen and water-vapor bands. The following characteristics were taken from the spectrograms.

1) The total width of the band at its base, t.

2) The vertical distance from the base of the band to its vertex (depth of band), l.

3) The width of the band at a distance of one-half of its depth from the vertex (half-width of band), Δ.

4) The ordinates of the continuous spectrum to the right and left of the band, and also the ordinates of the envelope corresponding to the center of the band.

The amount of radiant energy absorbed was proportional to the area of the band on the spectrogram. Since the outlines of the bands on the spectrograms were almost triangular, the area was approximately equal to the depth of the band multiplied by its half-width $l\Delta$. The

---

*Such a detector was first used to study weak luminescence at the Institute of Atmospheric Physics.

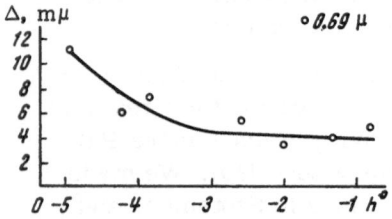

Fig. 2. Half-width of oxygen absorption band (0.69 μ) versus height of sun. Crimean Astrophysical Observatory, 4 October 1963.

total intensity of the twilight sky in the absence of an absorption band was proportional to the rectangle formed by I and t, i.e., proportional to the product It.

The relative absorption was determined as p = $l\Delta$/It.

The dependences of Δ, l/I (relative depth of band), and P upon the angle h° were studied. It was found that the half-widths had the greater dependence upon h°.

As examples, Figs. 1 and 2 show the band half-width Δ as a function of h° for two days.

It can be seen from the figures that as the sun sinks below the horizon, the half-width is increased (bands are spread out), which is evidently due to the increase in the thickness of the absorbing material on the beam path. Similar curves were obtained for the other bands. The spread of the points was greater on the graphs for the water-vapor bands. It is interesting to compare the relative absorptions p = $l\Delta$/It in the spectrum of the twilight sky with similar values in ground-level spectra of direct sunlight. From the comparison we can estimate the mass penetrated by the rays under twilight conditions by using any analytic relationship of the integral absorption in the band to the mass of the absorbing material. We used the square-root law in the form

$$A = C \sqrt{m}, \tag{1}$$

where m is the mass of absorbing material and C is a constant. The constant C for the respective bands was found from the spectra of direct sunlight in the near infrared region obtained by Faul [21]. The mass penetrated by the rays under twilight conditions was estimated by formula (1) using average (for several observation days) values of the relative absorption p = $l\Delta$/It, h° = $-30°$, and C values obtained from [21]. For the oxygen bands, we estimated the mass m, and for the water-vapor bands, we estimated the amount of water vapor W in centimeters of precipitation. Some of the results are given below.

Absorbing material..	O_2	O_2	H_2O	H_2O	H_2O
Band center, μ.....	0.69	0.76	0.72	0.82	0.93
m or W	21.5	14.4	30.4	57	52

The obtained atmospheric masses can exist in reality. Hence, we may conclude that the square-root law can, if only approximately, be used to describe the integral absorption in the oxygen and water-vapor bands.

To determine how the spectra varied with zenith angle, recordings were made not only at the zenith but also at zenith angles z = 40 and 60°. It was established that in general outline the spectra at z = 40 and 60° resembled the spectra at the zenith. The deepening of the absorption bands was considerable. On the average, the relative depth for z = 60° was greater than the zenith value by a factor of 1.5 for 0.69 μ and a factor of 1.2 for 0.72 μ.

To study the color variation of the twilight sky in the infrared region outside of the absorption bands of atmospheric gases, we found the dependence of the intensity ratio for two wavelengths $I(\lambda_2)/I(\lambda_1)$ upon the angle h°. Intensity $I(\lambda_2)$ corresponds to the higher wavelength. This ratio had an extremal variation in most cases. At the beginning of evening twilight, up to sun angles of -3 or $-4°$, the ratio $I(\lambda_2)/I(\lambda_1)$ increases, i.e., reddening of the twilight sky occurs. When the sun is lower below the horizon, the reddening is replaced by bluing, i.e., the ratio $I(\lambda_2)/I(\lambda_1)$ decreases. This extremal variation can be explained on the basis of G. V.

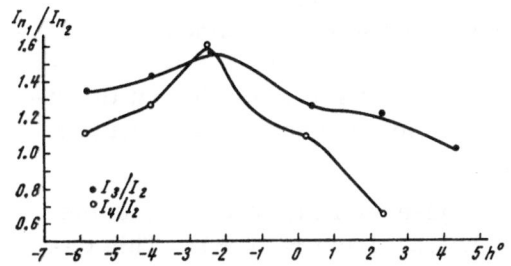

Fig. 3. Ratio I_{n_1}/I_{n_2} versus sun height $h°$. Crimean Astrophysical Observatory, 13 September 1963 (morning).

Rozenberg's theory of twilight. The brightness of the twilight sky can be expressed as

$$B_\lambda = I_0\omega_0 p^m m\tau_\lambda' (\overline{H}),$$

where I_0 is the intensity of direct solar radiation beyond the limits of the atmosphere; ω_0 the area of the solar disk expressed in circular measure; p the transmittance of the entire atmospheric thickness; m the mass of the atmosphere in the observation direction; $\tau_\lambda'(\overline{H})$ the optical thickness of the atmosphere above a certain level $\overline{H}$; and $\overline{H}$ the so-called effective shadow height. The occurrence of the above-mentioned effects is determined by the function $\overline{H}(\lambda)$. In the initial, unstable stage of the twilight, the effective shadow boundary is higher for shorter waves, outside the absorption bands of atmospheric gases, than it is for longer waves. Thus, the lower layers of the atmosphere are penetrated only by the longwave components of solar radiation. This leads to a relative predominance of longwave components in scattered sunlight (so-called reddening of the twilight sky). In other words, the reddening during the initial stage of twilight is explained by the Forbes effect. When the sun is 3 or 4° below the horizon, twilight becomes stabilized, and the so-called twilight ray is formed. From this moment, the effective shadow height $\overline{H}$ increases only due to an increase in the geometric shadow height H. The ratio $I(\lambda_2)/I(\lambda_1)$ reaches a maximum. When the sun is further below the horizon, according to theory there must be a relative intensification of the shortwave components in scattered sunlight, i.e., the twilight sky must become bluer, as has been observed (Fig. 3).

Starting with the moment of establishment of twilight, the color index of the twilight sky $CE_{\lambda_2}^{\lambda_1}$, which is defined as

$$CE_{\lambda_2}^{\lambda_1} = -1.08\log\frac{I(\lambda_1)}{I(\lambda_2)}$$

in accordance with [19, p. 282], must for the zenith be expressed by the formula

$$CE_{\lambda_2}^{\lambda_1} = C_{\lambda_2}^{\lambda_1} + 1.08\left[Q\Delta y_p - \frac{1}{2}\frac{dQ}{dh}(\Delta y_p)^2\right], \tag{2}$$

where Q is the logarithmic brightness gradient; Δy_p the difference between the perigees of rays with wavelengths λ_1 and λ_2 that illuminate the point of inflection of the atmospheric-transmittance curve; and $C_{\lambda_2}^{\lambda_1}$ a constant, which is independent of Δy_p (see [19] for details).

The value Δy_p is determined by the optical state of the atmosphere and is related to the extinction coefficient of the entire atmospheric thickness by the formula

$$\Delta y_p = y_{p\lambda_1} - y_{p\lambda_2} = \frac{1}{Q}\ln\frac{k_{\lambda_1}}{k_{\lambda_2}}.$$

If we use the Ångström formula $K = a\lambda^{-n}$, we obtain

$$\Delta y_p = \frac{1}{Q}\ln\frac{a\lambda_1^{-n}}{a\lambda_2^{-n}} = \frac{1}{Q}n\ln\frac{\lambda_2}{\lambda_1}. \tag{3}$$

We can find Δy_p from formula (2) by equating its right side to two experimentally obtained values of $CE_{\lambda_2}^{\lambda_1}$ taken at different arbitrarily selected $h°$. We can find n from formula (3).

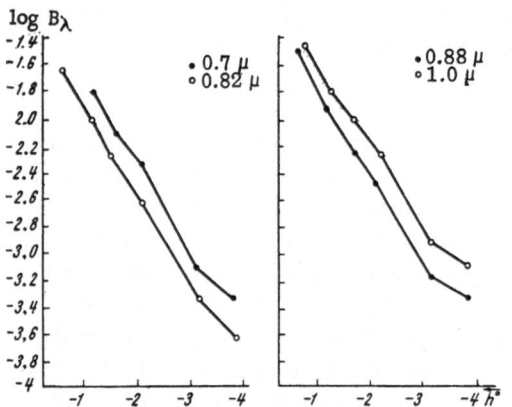

Fig. 4. Graphs of log $B_\lambda(h°)$ for 13 June 1963, Pulkovo.

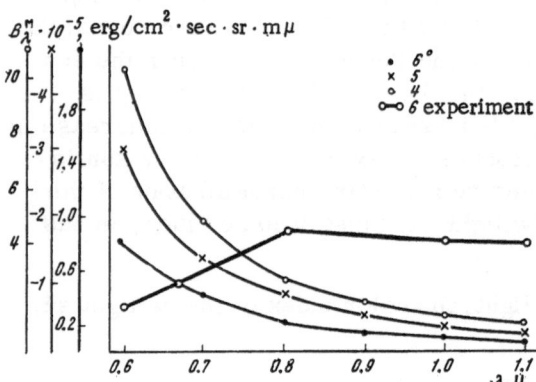

Fig. 5. Comparison of molecular component of twilight-sky brightness (calculation) with total brightness (experiment).

We calculated n by the above method for several twilight-sky spectra obtained in the Crimea. The mean value of n was 3.7, which was close enough to 4 for us to conclude that the atmosphere was almost molecular in the Crimea on those days.

Spectral observations were made in the range of sun heights from +8 to −6°. This enabled us to trace the brightness of the twilight sky B_λ in various spectral regions as a function of the sun's position relative to the horizon (angle h°).

Graphs of log $B_\lambda(h°)$ were plotted. Curves were plotted for the wavelengths corresponding to the centers of the absorption bands of oxygen and water vapor and for the maxima outside of the bands. Figure 4 shows example graphs of log $B_\lambda(h°)$ for 13 June 1963. An examination of the curves shows that the logarithm of the brightness decreases as the sun sinks below the horizon for all of the spectral regions studied, for the centers of the absorption bands as well as for the maxima outside of the bands. The curves for the centers of the bands are in general parallel to the curves for the maxima outside of the bands. This indicates the absence of a pronounced dependence of the relative depth of the band upon the sun's position in the range of +8 to −6°. The bands in the curves are due to changes in atmospheric conditions.

As we have already mentioned, we measured the spectral brightness of the twilight sky in absolute units. It is interesting to compare, if only roughly, the experimental data with the data calculated for molecular scattering. Since light is scattered in the atmosphere by air molecules (Rayleigh scattering) and by various suspended impurities (aerosols), the brightness formula for unstable twilight given above, $B_\lambda = I_0 \omega p^m m \tau'_\lambda (\overline{H})$, should be rewritten as

$$B_\lambda = I_0 \omega_0 p_m^m (\tau'^M_\lambda + \tau'^a_\lambda) = B_\lambda^M + B_\lambda^a, \qquad (4)$$

where τ^M_λ and τ^a_λ are the molecular and aerosol optical thicknesses and B_λ^M and B_λ^a are the molecular and aerosol components of the brightness. We can estimate B_λ^a by calculating B_λ^M and subtracting it from the total brightness. We calculated B_λ^M for sun angles of −4, −5, and −6°, took τ^M_λ for molecular scattering from Pendorf's data [22], and calculated $\overline{H}$ by the formula $\overline{H} = H + y_p$. Here, H is the geometric shadow height, which is given by the formula

$$H = \frac{R(1 - \sin \zeta)}{\sin \zeta},$$

where ζ is the solar zenith distance and R is the radius of the earth.

TABLE 1. Molecular Component
of Twilight-Sky Brightness
(in erg/cm² · sec · sr · mμ)

λ, μ	$h° = -6$	$h° = -5$	$h° = -4$
0.6	$0.79 \cdot 10^{-5}$	$3 \cdot 10^{-5}$	$1.04 \cdot 10^{-4}$
0.7	$0.4 \cdot 10^{-5}$	$1.35 \cdot 10^{-5}$	$4.85 \cdot 10^{-5}$
0.8	$0.19 \cdot 10^{-5}$	$8.4 \cdot 10^{-6}$	$2.6 \cdot 10^{-5}$
0.9	$0.15 \cdot 10^{-5}$	$5.3 \cdot 10^{-6}$	$1.8 \cdot 10^{-5}$
1.0	$0.89 \cdot 10^{-6}$	$3.3 \cdot 10^{-6}$	$1.2 \cdot 10^{-5}$
1.1	$0.55 \cdot 10^{-6}$	$2.1 \cdot 10^{-6}$	$0.77 \cdot 10^{-6}$

TABLE 2. Aerosol Component of Twilight-
Sky Brightness

λ, μ	B_λ^a, erg/cm²·sec·sr·mμ	B_λ^a/B_λ^M	τ_λ^a
0.6	$0.3 \cdot 10^{-5}$	0.2	$1.3 \cdot 10^{-2}$
0.7	$1.6 \cdot 10^{-5}$	1.7	$6.1 \cdot 10^{-2}$
0.8	$4 \cdot 10^{-5}$	7.0	$1.5 \cdot 10^{-1}$
0.9	$3.9 \cdot 10^{-5}$	12.0	$1.4 \cdot 10^{-1}$
1.0	$3.7 \cdot 10^{-5}$	11.5	$8 \cdot 10^{-2}$

To determine y_p, we used the expression $y_p = 12.5(1 - 2 \log \lambda)$ km for an aerosol atmosphere ([19], p. 218). The results of the calculations of brightness for molecular scattering B_λ^M as a function of wavelength for $h° = -6, -5$, and -4 are given in Table 1 and Fig. 5.

As is apparent from Fig. 5, B_λ^M decreases rapidly with an increase in wavelength. The ratio of the brightness at 0.6 μ to the brightness at 1 μ is $B_{\lambda = 0.6\mu}^m / B_{\lambda = 1\mu}^m = 8.9$. The heavy line in Fig. 5 represents the curve of spectral brightness plotted from average data for Pulkovo at $h° = -6°$. It can be seen that the difference between the measured and calculated brightness increases with wavelength.

To assess the aerosol component of the twilight-sky brightness, we found the differences between the measured and calculated values. Table 2 shows the obtained aerosol components B_λ^a for a number of wavelengths, and also the ratios of the aerosol component to the molecular B_λ^a/B_λ^M. Table 2 also gives the aerosol optical thicknesses τ_λ^a as calculated by our formula (4), on the basis of B_λ^a values, and reduced to sea level. The maximum of τ_λ^a fell within the wavelength range of 0.8-0.9 μ.

Summary

Infrared spectra of the twilight sky in absolute units were obtained for the first time, using photoelectric photometry, for two geographical points.

Certain regularities were discovered in the oxygen and water-vapor absorption bands with centers at 0.63, 0.72, 0.76, and 0.87 μ in the near infrared region. It was established that the absorption bands are spread out as the sun sinks below the horizon. The atmospheric mass penetrated by rays under twilight conditions when the sun is 3° below the horizon was estimated.

Effects connected with the color variation of the twilight sky in the near infrared region were studied. It was found that reddening occurs as a result of selective scattering at the beginning of twilight, up to $h° = -3$ to $-4°$. After this, the reddening is replaced by bluing.

The contributions of the molecular and aerosol components to the brightness of the twilight sky were evaluated. The ratio $\tau_\lambda^a/\tau_\lambda^m$ increases from 0.2 to 11.5 as wavelength increases from 0.6 to 1 μ.

The dependence of brightness upon the position of the sun below the horizon inside and outside of the absorption bands, and also the dependence of the depth of the absorption bands upon the observation angle on the solar vertical circle were studied.

Further spectrophotometric investigation of twilight with better apparatus could provide spectra when the sun is more than 6° below the horizon, up to nightfall, which would give new information on the structure of the earth's atmosphere.

I thank Professor S. F. Rodionov, in whose laboratory this work was performed.

LITERATURE CITED

1. V. G. Fesenkov, Trudy Glavnoi Rossiiskoi astrofiz. obs., Vol. 2, p. 7 (1923).
2. V. G. Fesenkov, Astron. zh., Vol. 7, No. 2, p. 100 (1936).
3. I. N. Yaroslavtsev, Geophys. and Meteorol., Vol. 5, No. 1, p. 1 (1928); Vol. 6, No. 2, p. 143 (1929).
4. F. Schembor, Gerl. Beitr. Geophys., Vol. 28, p. 279 (1930).
5. B. J. Rayleigh, Proc. Roy. Soc., Vol. A124, No. 795, p. 395 (1929).
6. Bauer, Donjon, and Langevin, Compt. rend., Vol. 178, p. 2115 (1924).
7. F. Link, J. Observateurs, Vol. 17, p. 161 (1934).
8. T. G. Megrelishvili, Doklady Akad. Nauk SSSR, Vol. 53, No. 2, p. 127 (1946).
9. T. G. Megrelishvili, Byull. Abastum. astrofiz. obs., No. 9, pp. 1-142 (1948).
10. T. G. Megrelishvili, Izv. Akad. Nauk SSSR, seriya geofiz., No. 8, p. 976 (1956).
11. T. G. Megrelishvili and I. A. Khvostikov, Doklady Akad. Nauk SSSR, Vol. 59, No. 7, p. 1283 (1948).
12. P. Grunner and Kleinert, Die Dämmerungerscheinungen, Hamburg (1927).
13. F. Volz and R. M. Goody, Blue Hill Meteorol. observ. Scient. Rept. No. 1, on contract AF-19-604 (4546), Harvard Univ. (1960).
14. N. B. Divari, Doklady Akad. Nauk SSSR, Vol. 122, No. 5, p. 795 (1958).
15. F. E. Volz and R. M. Goody, J. Atmosph. Sci., Vol. 19, No. 5, p. 385 (1962).
16. V. I. Cherneev and M. F. Vuks, Doklady Akad. Nauk SSSR, Vol. 14, p. 77 (1937).
17. J. Dufay, Gauzit, Ann. Astrophys., Vol. 9, p. 135 (1946).
18. A. Kh. Darchiya, Izv. GAO Akad. Nauk SSSR, No. 165 (1960).
19. G. V. Rozenberg, Twilight [in Russian], Fizmatgiz (1963) [English translation, Plenum Press, New York (1966)].
20. E. V. Gnilovskoi, Vestnik Leningr. univ., seriya fiz. i khim., No. 4, Issue 1 (1964).
21. Faul, Smithsonian Meteorological Tables, Washington (1951).
22. R. Pendorf, J. Opt. Soc. Am., Vol. 47, No. 2, p. 176 (1957).
23. E. D. Sholokhova and M. S. Frish, Doklady Akad. Nauk SSSR, Vol. 105, No. 6, p. 1218 (1956).

DYNAMICS OF THE BRIGHTNESS EXTREMES
OF THE TWILIGHT SKY

V. K. Pyldmaa

The brightness distribution of the twilight sky along the solar vertical circle contains a number of noteworthy extremal points whose zenith distance varies regularly with the solar zenith distance ξ, wavelength λ, the form of the scattering indicatrix $f(\varphi)$, the contribution of multiple scattering, and atmospheric transmittance p. We shall consider the essence of these variations on the basis of G. V. Rozenberg's theory of twilight [1]. The theoretical results will be compared with spectrophotometric data [2-4].

The following assumptions will be made.

1. During twilight, at any moment, the brightness of the sky consists of two components: $I = I_1 + I_2$ (I_1 is the brightness from primary scattering and I_2 is the brightness from multiple scattering). The relationship of these components is extremely variable. In many cases, the contribution of one of them is negligible in a first approximation.

The brightness of primary scattering is

$$I_1 = I_0 \omega_0 p^m m \tau,$$

where $I_0(\lambda)$ is the brightness of the sun; ω_0 the area of the solar disk; m(z) the mass of the atmosphere on the line of sight; τ the optical thickness for scattering above the level $\overline{H}$; $\overline{H}(\lambda, \xi, z, A)$ the effective shadow height; z the zenith distance of the line of sight; and A the azimuth of the line of sight.

2. Radiation absorption is negligible.

3. The coordinate origin is placed such that z is positive on the solar vertical circle and negative on the antisolar vertical circle.

4. $\frac{\partial}{\partial z}(mp^m)|_{z=\pm z_1} = 0.$

5. The twilight phase covering $\xi = 90-100°$ is considered.

Analysis of our results allow the following conclusions to be drawn.

1. In the solar vertical circle region $\pi/2 \geq z > z_1$ there is a brightness maximum whose zenith distance increases with an increase in λ and ξ. Elongation of the scattering indicatrix brings this maximum even closer to the horizon.

2. Brightness extremes cannot exist when $0 < z \leq z_1$.

3. At the moment of sunset, the brightness minimum is on the antisolar vertical circle in the immediate vicinity of the zenith. It moves away from the zenith as ξ increases and as

λ decreases. This continues as long as the role of multiple scattering is negligible. Taking into account the indicatrix effect shows that when the minimum of the function $p^m m\tau$ is closer to the zenith than the minimum of the indicatrix (the scattering angle $\varphi = \xi - z$), then, under the influence of the indicatrix, the actual brightness minimum of primary scattering moves away from the zenith; otherwise, it approaches it.

As the sun sinks, the role of multiple scattering increases, and after a time, it can no longer be ignored. It is known that on the antisolar vertical circle the brightness of multiple scattering increases in the direction of the horizon. In the immediate vicinity of the horizon, the increased extinction causes it to decrease, i.e.,

$$\frac{\partial I_2}{\partial z} \begin{cases} < 0, & \text{if } \; 0 > z > z_2, \\ > 0, & \text{if } \; -\frac{\pi}{2} \leqslant z < z_2. \end{cases}$$

Therefore, the effect of multiple scattering on the zenith distance of the circumzenithal brightness minimum is to cause the minimum to approach the zenith.

It follows from the above that, in the region of the sky in question, multiple scattering plays a minor role from sunset up to solar zenith distances at which the brightness minimum stops withdrawing from and begins approaching the zenith. According to spectrophotometric data, these solar zenith distances are $\xi \approx 96\text{-}97°$. At greater solar zenith distances, the dominant role is gradually taken over by higher-order scattering.

4. Let the circumzenithal brightness minimum have zenith distance $z = z_3$. Then the brightness maximum of primary scattering must be found at $-z_1 < z < z_3$. It approaches the zenith as ξ increases. On the other hand, it moves away from the zenith as λ increases. This latter movement continues up to a certain ξ, after which a reverse spectral variation occurs. This is due to the growing effect of multiple scattering. Since the relative contribution of multiple scattering increases as λ decreases, its effect manifests itself sooner in the shortwave region, causing this reverse spectral variation.

5. Let us assume that the maximum on the antisolar vertical circle has zenith distance $z = z_4$. At least one brightness extreme must be found in the zenith distance range $z_2 \leq z \leq z_4$. Spectrophotometric data show that usually one or two maxima are observed here. The existance of two maxima should be considered common, and the disappearance of one of them should be attributed to a decrease in the contribution of either primary or multiple scattering.

According to photometric data, in the region of the sky in question two brightness maxima can exist simultaneously only for a short period of time ($\xi = 94\text{-}96°$). Therefore, primary and multiple scattering play basically identical roles in this twilight phase in this region of the sky. When $\xi < 94°$, the brightness of multiple scattering is negligible in comparison with that of primary scattering, and when $\xi > 96°$, multiple scattering dominates.

Elongation of the scattering indicatrix causes these maxima to approach the horizon.

LITERATURE CITED

1. G. V. Rozenberg, Twilight [in Russian], Fizmatgiz (1963) [English translation, Plenum Press, New York (1966)].
2. V. K. Pyldmaa, Izv. Akad. Nauk Est.SSR, seriya fiz.-matem. i tekhn. nauk, Vol. 13, No. 3, p. 192 (1964).
3. V. K. Pyldmaa, in: Studies of the Radiation Conditions of the Atmosphere [in Russian], Tartu (1967).
4. V. K. Pyldmaa, Izv. Akad. Nauk SSSR, seriya fizika atmosfery i okeana, Vol. 1, No. 11, p. 1168 (1965).

DETERMINING ATMOSPHERIC TRANSMITTANCE FROM THE BRIGHTNESS OF THE TWILIGHT SKY

V. K. Pyldmaa

In order to interpret data on photometry of the twilight sky, one must know the spectral transmittance of the atmosphere. During twilight, however, it cannot be determined with the aid of either the sun, which is already below the horizon, or the stars, which are not yet bright enough against the background of the twilight sky. G. V. Rozenberg [1] has proposed a method of determining atmospheric transmittance from the brightness distribution of the twilight sky on the meridian perpendicular to the solar vertical circle.

If we assume that

$$I_1 = I_0 \omega_0 p^m m \tau \tag{1}$$

and

$$I_2 = I_0 \omega_0 p^m m b, \tag{2}$$

where b is slightly dependent upon the viewing direction, we obtain

$$\tau^* = \frac{\ln m + \ln I\,(\xi,\,\lambda)_{z=0} - \ln I\,(\xi,\,\lambda,\,z)_{A=\pm\frac{\pi}{2}}}{m-1}, \tag{3}$$

where τ^* is the optical thickness of the atmosphere from the earth's surface to the upper limit of the atmosphere; τ the optical thickness for scattering above the effective shadow height $\bar{H}$; I_1 and I_2 the brightnesses from primary and multiple scattering; I_0 the brightness of the sun; ω_0 the area of the solar disk; p the vertical atmospheric transmittance; m the atmospheric mass in the viewing direction z; A the azimuth of the viewing direction; and ξ the solar zenith distance.

Transmittance was found by formula (3) for a number of twilights using spectrophotometric data [2]. The results showed that G. V. Rozenberg's method of determining spectral transmittance can be recommended for $30° \le z \le 60°$ and $90° \le \xi \lessapprox 96°$, and also, in all probability, when $\xi > 100°$.

When $\xi < 96°$, the role of multiple scattering is negligible. When $96° < \xi < 100°$, the contribution of multiple scattering is already appreciable, but since the range of heights that is chiefly responsible for multiple scattering at small ξ is situated in the lowest layers of the atmosphere and is shifted upward as the sun sinks below the horizon, being detached from the earth's surface only when $\xi = 98$–$99°$ [3, 4], assumption (2) is not valid at $\xi < 100°$.

LITERATURE CITED

1. G. V. Rozenberg, Twilight [in Russian], Fizmatgiz (1963) [English translation, Plenum Press, New York (1966)].
2. V. K. Pyldmaa, Izv. Akad. Nauk Est. SSR, Seriya Fiz.-matem. i tekhn. nauk, Vol. 13, No. 3, p. 192 (1964).
3. J. V. Dave, Proc. Ind. Acad. Sci., Vol. 43, Section A, No. 6, p. 336 (1956).
4. V. K. Pyldmaa, Izv. Akad. Nauk SSSR, seriya fizika atmosfery i okeana, Vol. 1, No. 11, p. 1168 (1965).

COMPARISON OF SOME METHODS OF ELIMINATING THE EFFECT OF THE EARTH'S ATMOSPHERE FROM THE OBSERVED BRIGHTNESS OF ZODIACAL LIGHT

N. B. Divari, N. I. Komarnitskaya, and S. N. Krylova

The proper allowance for tropospheric scattering is very important in reducing observed brightnesses of zodiacal light. This essential question was first raised by V. G. Fesenkov and examined in detail by him in [1-4]. According to Fesenkov's theory, the relationship between the observed brightness of zodiacal light J_{app} (free of night airglow and the stellar background) and the extra-atmospheric brightness J is

$$J = K J_{app} p^{-\sec z_0},$$ (1)

where p is the transmission coefficient of the earth's atmosphere. The factor K takes into account tropospheric scattering of zodiacal light and is determined by the formula

$$K = \left(1 + \frac{B(z_0, A_0)}{J(z_0, A_0) p^{\sec z_0}}\right)^{-1},$$ (2)

where $B(z_0, A_0)$ is the tropospheric component of the brightness of zodiacal light (i.e., the intensity of zodiacal light scattered by the troposphere) at a point with zenith distance z_0 and azimuth A_0. It is obvious that

$$J_{app} = J(z_0, A_0) p^{\sec z_0} + B(z_0, A_0).$$ (3)

At small solar dip angles g < 20°, twilight glow must also be taken into account, as has been shown in [5]. The twilight component $S(g, z_0, A_0)$ must be eliminated from the value J_{app} in the above formulas.

The extra-atmospheric brightnesses of zodiacal light that have been reported by most researchers were obtained without allowance for the twilight component. Moreover, the tropospheric component has frequently been taken into account approximately by increasing the transmission coefficient by 0.02, so that the relationship between J and J_{app} was expressed as

$$J = J_{app} (p + 0.02)^{-\sec z_0}.$$ (4)

As V. G. Fesenkov has shown [1], this formula does not meet the accuracy of modern measurements.

In view of this, it is interesting to reconsider the earlier published extra-atmospheric brightnesses of zodiacal light and to compare the results of various methods of eliminating

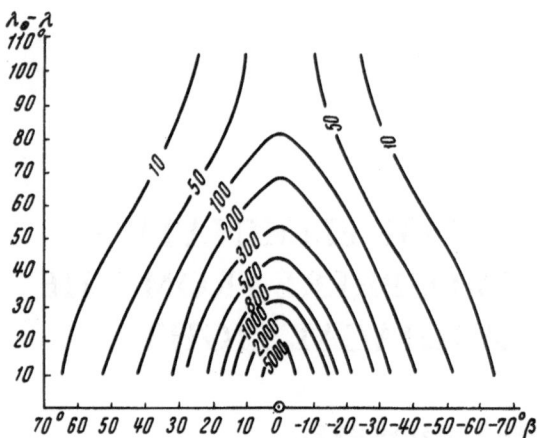

Fig. 1. Isophots of zodiacal light taken as first approximation for calculating tropospheric scattering.

the tropospheric component. For this, we used photoelectric observations of zodiacal light made in Egypt in the autumn of 1957 [6, 7].

The apparent brightness of zodiacal light J_{app} from which had been eliminated airglow, the stellar background, and twilight glow were corrected for tropospheric scattering by V. G. Fesenkov's method. The values of the factor K were taken from the tables in [3]. The extra-atmospheric brightnesses of zodiacal light were obtained from observations made with an interference filter (λ_{max} 522 mμ) on three nights and a green glass filter (λ_{max} 541 mμ) on four nights. From these values, we found the isophots and extrapolated them to the required part of the celestial sphere. The thus-obtained brightnesses were used as a first approximation for the extra-atmospheric brightnesses of zodiacal light. Then these brightnesses were used to calculate tropospheric scattering for the observations of zodiacal light. The tropospheric component was calculated because, as V. G. Fesenkov points out [3], his tables for K, which takes into account tropospheric scattering, are approximate, since purely theoretical isophots of zodiacal light were used as the initial isophots. The form of the initial isophots has a great effect on the values of K, especially at the boundaries of the zodiacal cone.

The brightnesses (first approximation) that we used to calculate tropospheric scattering are shown in the form of isophots in Fig. 1. These brightnesses are expressed in absolute units: in the number of stars of magnitude 10 per square degree. For the region of the celestial sphere not included in Fig. 1, the brightnesses were assumed to be zero. Thus, we calculated the illumination from only the zodiacal cones. It must be noted that the isophots of extra-atmospheric zodiacal light that we obtained from observations and used as a first approximation correspond to brightnesses that are somewhat lower than the mean brightnesses in [8] from various researchers.

The tropospheric component B(z_0, A_0) was calculated on a Ural-2 computer by V. G. Fesenkov's formula

$$B(z_0, A_0) = Cm_0 \iint J(z, A) f(\vartheta) \varphi(z, z_0) \, dz \, dA, \qquad (5)$$

where $f(\vartheta)$ is the scattering indicatrix, which was used, in accordance with E. V. Pyaskovskaya-Fesenkova's observations [9], in the form

$$Cf(\vartheta) = \frac{1}{17.4} [1 + 5(e^{-3\vartheta} - 0.009) + 0.55 \cos^2 \vartheta], \qquad (6)$$

where ϑ is the scattering angle, m the atmospheric mass, and

$$\varphi(z, z_0) = \frac{p^m - p^{m_*}}{m_0 - m}, \quad \text{if } z \neq z_0 \ (m \neq m_0),$$
$$\varphi(z, z_0) = -p^m \ln p, \quad \text{if } z = z_0 \ (m = m_0). \qquad (7)$$

For the transmission coefficient we used p = 0.83, which is its mean value for the observations with the green filter (p = 0.83) and is close to its mean value for the observations with the interference filter λ 522 (p = 0.82). Table 1 gives for three almucantars z = 80°, z = 70°, and

TABLE 1. Brightness of Tropospheric Component for i = 90°
(expressed in number of stars of magnitude 10 per square degree)

Azimuth A	Solar dip angle						
	16°	20°	24°	28°	32°	36°	40°
z = 80°							
0°	40	29	22	18	15	12	7
10	36	27	21	17	14	11	7
20	31	23	18	15	12	10	6
30	25	20	15	13	10	9	5
40	21	17	13	11	9	7	4
50	18	14	11	9	7	6	3
60	15	12	9	8	6	5	3
70	14	11	8	7	6	5	3
80	12	10	8	6	5	4	2
110	12	9	7	6	5	4	2
140	13	10	8	6	5	4	2
170	14	11	9	7	6	5	2
z = 70°							
0°	26	19	15	12	10	8	7
10	24	18	14	12	10	8	7
20	21	16	13	10	9	7	6
30	18	13	11	9	7	6	5
40	15	10	9	7	6	5	4
50	13	9	8	6	5	4	3
60	11	8	7	5	4	4	3
70	10	7	6	5	4	3	3
80	10	7	5	4	4	3	2
110	8	6	5	4	3	3	2
140	8	7	5	4	3	3	2
170	9	7	5	4	4	3	2
z = 60°							
0°	18	14	11	9	7	6	5
10	17	13	10	8	7	6	5
20	15	12	10	8	6	5	4
30	13	10	8	7	5	5	4
40	11	9	7	5	5	4	3
50	10	8	6	5	4	3	3
60	8	7	5	4	3	3	2
70	7	6	5	4	3	3	2
80	7	5	4	3	3	2	2
110	6	5	4	3	2	2	2
140	6	5	4	3	2	2	2
170	7	5	4	3	3	2	2

z = 60° the calculated brightnesses of the tropospheric component $B(z_0, A_0)$ for solar dip angles of from 16 to 40° for the case when the ecliptic is perpendicular to the horizon (i = 90°). The azimuth A is read from the solar vertical circle. The brightnesses in Table 1 are expressed in absolute units: the number of stars of magnitude 10 per square degree. In order to compare our results with the results of V. G. Fesenkov's calculations [3], we found from our data the values of K as determined by formula (2). Our K values near the ecliptic turned out to be rather close to those of V. G. Fesenkov, but far from the ecliptic our values were appreciably lower. This was because we calculated the tropospheric scattering of only the zodiacal cone and assumed that there was zero zodiacal light at a distance from the ecliptic. This caused K to decrease rapidly as azimuth increased. At high azimuths, K was equal to zero. As an example, our K values are compared with Fesenkov's for the almucantar z = 70° at solar dip angle g = 20° and ecliptic inclination from the horizon i = 90°:

A	From Fesenkov [3]	Our calculation	A	From Fesenkov [3]	Our calculation
0°	0.949	0.951	60°	0.647	0,520
10	0.900	0.933	70	0.621	0,320
20	0.841	0.888	80	0.605	0
30	0.783	0.824	110	0.578	0
40	0.728	0.757	140	0.617	0
50	0.682	0.520	170	0.706	0

As can be seen, K is greatly dependent upon the zodiacal brightness used in the calculation. This is especially true far from the ecliptic. But inasmuch as the corrections for tropospheric scattering themselves are small in comparison with the zodiacal-light brightnesses, it is obvious that in calculating the brightness of the tropospheric component in absolute units, the discrepancies in the zodiacal-light isophots that are used will not be so great. With this in mind, we reduced the observed brightnesses of zodiacal light for tropospheric scattering using the tropospheric-component brightnesses in absolute units that we had calculated.

In this case, the extra-atmospheric brightness of zodiacal light J is determined by

$$J(z_0, A_0) = [J_{app} — B(z_0, A_0) — S(g, z_0, A)] \, p^{-sec \, z_*} \tag{8}$$

when twilight glow is taken into account, or by

$$J(z_0, A_0) = [J_{app} — B(z_0, A_0)] \, p^{-sec \, z_*} \tag{9}$$

when twilight glow is ignored.

The brightnesses obtained by formula (8) were used as the final extra-atmospheric brightnesses. These brightnesses were calculated taking into account the twilight component and are presented in each first row (brightness J_1) of Table 2 for the observations with the interference filter λ 522 and of Table 3 for the observations with the green filter. To compare various methods of allowing for the effect of the atmosphere on the extra-atmospheric brightnesses of zodiacal light, the tables show the results obtained, without taking into account twilight glow, using the following three methods: formula (9) (second rows, J_2); formula (1) using Fesenkov's K values (third rows, J_3); and approximation formula (4) (fourth rows, J_4).

The effect of the twilight component can be evaluated by comparing the J_1 and J_2 values in the first and second rows. It was found to be considerable at small elongations, i.e., for observations near the horizon, although we used observations that had been made only up to the beginning of astronomical twilight (g > 17.°5). For observations at smaller solar dip angles, the effect of twilight is considerably greater. Comparison of J_2 and J_3 shows that our corrections for tropospheric scattering are somewhat lower than those obtained using V. G. Fesenkov's K factors [3]. This must be a result of the difference between the zodiacal-light brightnesses used to calculate tropospheric scattering. These brightnesses are not given in [3], but it is indicated that theoretical isophots of zodiacal light were used for the calculations. We used isophots obtained from observations, which are shown in Fig. 1.

Brightnesses J_4, which were found using approximation formula (4) and are given in each fourth row of Tables 2 and 3, were somewhat lower near the ecliptic and somewhat greater at a distance from the ecliptic than were brightnesses J_1, which were obtained taking into account tropospheric scattering and twilight. This is illustrated in Fig. 2, which shows the brightness of extra-atmospheric zodiacal light as a function of ecliptic latitude β for a constant longitude relative to the sun ε = $λ_⊙$ − λ = 35°. When the dependence of brightness upon ecliptic latitude β is approximated by the function J = J_m exp[−k $(β − β_m)^2$], where $β_m$ is the ecliptic latitude at which the brightness along cross section ε = const reaches a maximum, then for the factor K we obtain average values for all of the cross sections ε = const examined (Table 4).

TABLE 2. Extra-Atmospheric Brightness of Zodiacal Light
Obtained by Four Methods of Allowance for Effect of Atmosphere
Interference Filter λ 522
(expressed in number of stars of magnitude 10 per square degree)

Ecliptic latitude	Bright-ness	Longitude relative to sun								
		35°	40°	45°	50°	5.°	60°	65°	70°	75°
0°	J_1	985	687	515	392	310	248	202	160	127
	J_2	1078	762	547	412	320	255	205	165	132
	J_3	989	697	500	380	297	238	188	153	124
	J_4	933	683	490	377	300	230	187	149	124
5	J_1	868	622	487	378	290	232	193	155	127
	J_2	998	670	513	402	300	240	195	158	132
	J_3	937	653	460	350	281	228	188	160	132
	J_4	887	637	467	360	285	233	186	154	123
−5	J_1	727	543	393	322	251	210	164	135	106
	J_2	830	600	457	329	266	213	168	143	118
	J_3	745	520	393	290	223	187	148	123	98
	J_4	783	570	412	315	228	190	156	127	103
10	J_1	607	447	362	292	235	190	154	125	108
	J_2	702	510	383	304	242	195	156	127	105
	J_3	555	443	343	265	217	175	149	133	108
	J_4	627	480	363	292	233	187	158	132	102
−10	J_1	503	393	302	241	188	155	125	102	76
	J_2	618	435	323	251	193	157	127	101	82
	J_3	533	377	289	220	170	142	115	90	72
	J_4	585	410	320	242	193	155	125	97	78
20	J_1	268	218	178	145	120	98	77	64	50
	J_2	323	253	192	150	125	100	80	65	52
	J_3	230	215	167	135	110	92	78	63	48
	J_4	302	238	185	158	127	100	85	68	49
−20	J_1	238	188	152	122	95	73	58	47	38
	J_2	327	233	173	127	100	76	62	54	41
	J_3	262	199	148	113	90	62	46	38	25
	J_4	322	245	178	133	103	78	58	42	28
30	J_1	132	111	86	65	51	41	30	26	21
	J_2	172	124	88	70	55	43	35	31	25
	J_3	132	105	83	67	58	37	29	21	15
	J_4	193	145	103	89	66	48	37	27	17
−30	J_1	98	82	66	51	35	26	20	14	10
	J_2	160	123	87	59	43	30	21	16	14
	J_3	113	89	67	49	34	25	16	10	8
	J_4	164	135	102	65	44	29	21	12	7

TABLE 3. Extra-Atmospheric Brightnesses of Zodiacal Light
Obtained by Four Methods of Allowance for Effect of Atmosphere
Green Filter
(expressed in number of stars of magnitude 10 per square degree)

Ecliptic latitude	Bright-ness	Longitude relative to sun								
		35°	40°	45°	50°	55°	60°	65°	70°	75°
0°	J_1	900	612	460	366	297	218	190	150	129
	J_2	1030	740	562	428	317	244	207	172	145
	J_3	952	705	522	397	310	257	212	177	143
	J_4	855	660	475	365	286	229	189	159	130
5	J_1	780	547	421	321	262	210	172	144	132
	J_2	932	615	472	380	295	232	196	163	140
	J_3	745	571	447	345	281	229	194	165	128
	J_4	762	570	429	335	256	209	172	149	120
−5	J_1	740	510	391	297	234	182	159	131	110
	J_2	865	606	469	365	284	217	182	155	131
	J_3	790	552	415	320	264	223	177	146	115
	J_4	738	527	417	315	252	197	164	138	110
10	J_1	540	397	301	242	215	172	141	116	99
	J_2	634	484	373	296	235	194	145	131	111
	J_3	521	414	330	266	222	180	155	132	103
	J_4	572	439	344	266	210	173	143	122	99
−10	J_1	527	350	259	205	160	131	111	98	78
	J_2	619	432	334	259	212	171	142	122	104
	J_3	535	387	302	237	197	167	137	109	92
	J_4	540	400	306	239	191	152	129	106	86
20	J_1	272	204	155	109	100	85	69	62	55
	J_2	340	255	200	161	125	114	92	75	64
	J_3	282	210	162	136	111	92	75	64	53
	J_4	304	235	182	146	116	99	79	66	56
−20	J_1	241	160	118	97	77	63	47	42	38
	J_2	284	214	172	131	107	91	72	61	52
	J_3	222	171	145	117	96	83	71	54	29
	J_4	240	191	152	122	102	85	68	58	43
30	J_1	147	109	83	65	51	40	33	27	27
	J_2	192	140	112	85	74	60	45	40	29
	J_3	155	102	87	67	56	45	37	31	13
	J_4	185	135	110	87	69	54	41	32	30
−30	J_1	83	49	39	28	20	19	18	15	14
	J_2	99	72	55	41	38	32	26	23	20
	J_3	90	42	37	29	20	23	20	18	16
	J_4	100	76	57	46	44	42	40	32	27

TABLE 4. K Factors Determined from Brightnesses J_1, J_2, and J_4 for Two Filters

Method	K	
	Fliter λ 522	Green filter
Brightness J_1 (allowance for twilight and tropospheric scattering)	0.0040	0.0035
Brightness J_2 (without allowance for twilight)................	0.0037	0.0033
Brightness J_4 (approximation formula (4)).................	0.0033	0.0031

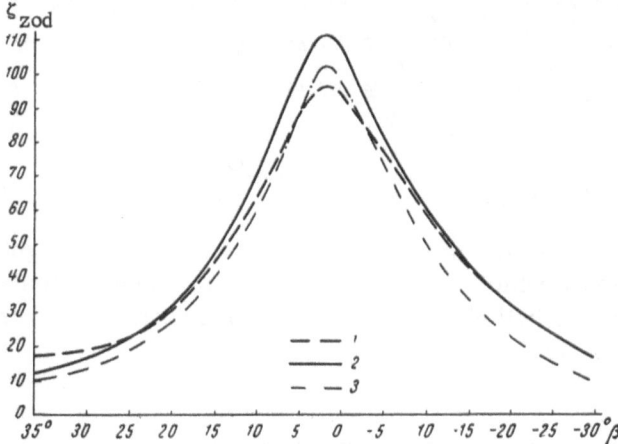

Fig. 2. Brightnesses J_1, J_2, and J_4 as functions of ecliptic latitude β along cross section $\varepsilon = 35°$: 1) p + 0.02; 2) with allowance for tropospheric scattering; 3) with allowance for tropospheric scattering and twilight.

As can be seen from Table 4, in the three cases in question, when $|\beta|$ increases, the brightnesses along cross sections ε = const decrease more rapidly for the observations with the interference filter λ 522 than for the observations with the green glass filter. But for both the interference filter and the glass filter, the brightness along cross section ε = const decreases most rapidly when tropospheric scattering and twilight glow are taken into account. When approximation formula (4) is used and twilight glow is ignored, the decrease in brightness with an increase in $|\beta|$ is appreciably slower. The brightnesses given by formula (4) also decrease more slowly with distance from the ecliptic than the brightnesses obtained from formula (9) (ignoring twilight). V. G. Fesenkov obtained this result in [4]. When twilight glow is taken into account, the brightness of zodiacal light increases somewhat more slowly as elongation decreases than when twilight is ignored. All of this is due to the fact that the isophots of zodiacal light plotted from brightnesses J_1, from which tropospheric scattering and twilight glow have been deducted, are narrower than isophots plotted from brightnesses containing twilight glow. The difference in the degree of narrowing of the isophots can be seen in Figs. 3, 4, and 5, which show isophots of zodiacal cones plotted from brightnesses J_1, J_2, and J_4 for observations with the interference filter λ 522. As can be seen, there are differences between the isophots, but they are not very great. The differences would have been considerably greater if observations made at smaller solar dip angles had been used. The difference in the

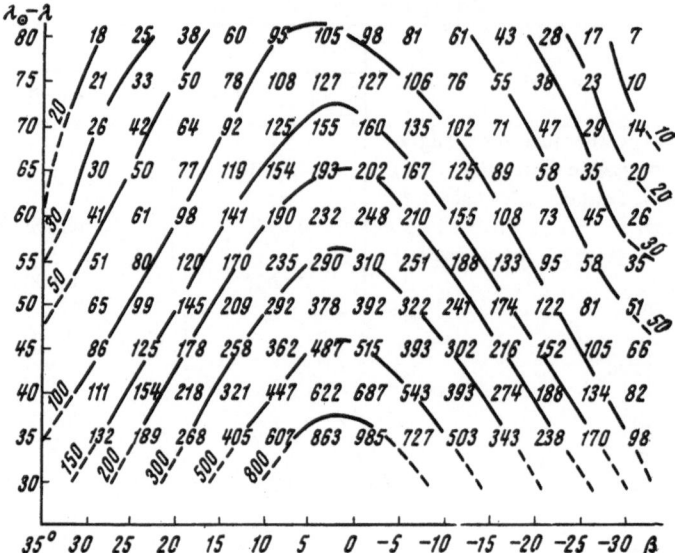

Fig. 3. Isophots of zodiacal light plotted from brightnesses J_1, which take into account tropospheric scattering and twilight glow (observations with interference filter λ 522).

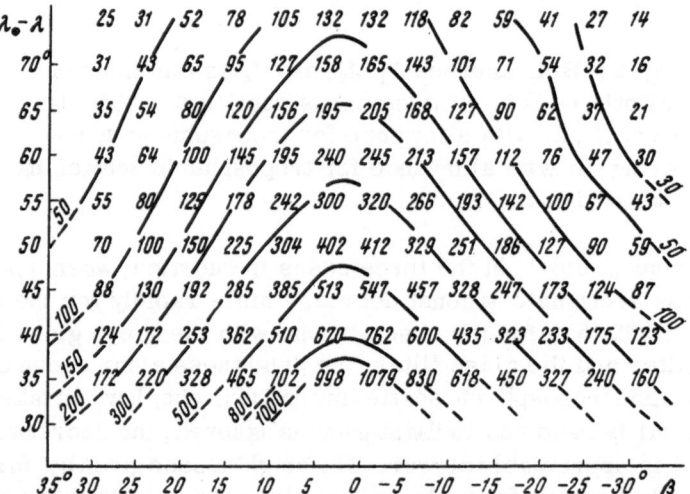

Fig. 4. Isophots of zodiacal light plotted from brightnesses J_2, taking into account tropospheric scattering by formula (8) but ignoring twilight glow (observations with interference filter λ 522).

isophots is smaller for the observations with the green glass filter than for the observations with the interference filter.

Our comparison of the various methods of allowing for the effect of the atmosphere on the brightness of zodiacal light shows that, to obtain true extra-atmospheric brightnesses of zodiacal light, twilight must be taken into account and reduction for tropospheric scattering

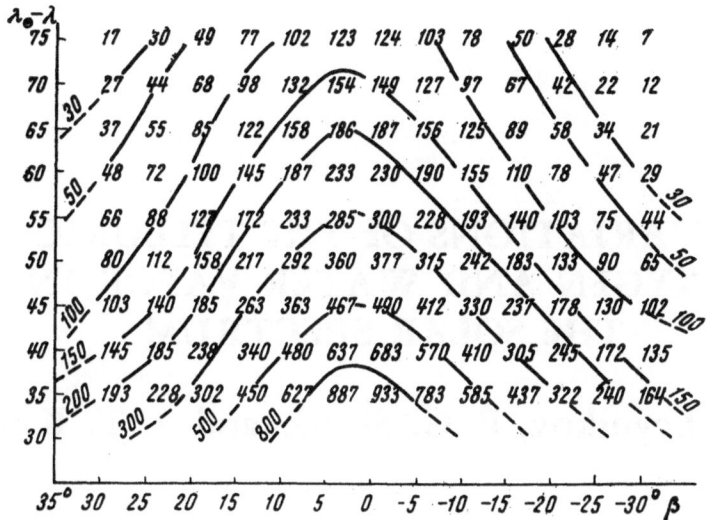

Fig. 5. Isophots of zodiacal light plotted from brightnesses J_4, ignoring twilight glow but taking into account tropospheric scattering by increasing transmission coefficient by 0.02 in accordance with approximation formula (4) (observations with interference filter λ 522).

must be done not by approximation formula (4) but by more exact formula (8). It must be borne in mind that tropospheric scattering as calculated by formula (5) is very greatly dependent upon the transmission coefficient of the earth's atmosphere. The correction for tropospheric scattering, therefore, must be calculated in each specific case for the corresponding transmission coefficient. Since the brightness of the twilight sky can vary from night to night, the twilight component of zodiacal light must be determined immediately after or before each observation. To avoid large corrections for twilight glow, zodiacal light should be observed at solar dip angles greater than 18°, or even greater than 20°.

LITERATURE CITED

1. V. G. Fesenkov, Astron. zh., Vol. 35, p. 323 (1958).
2. V. G. Fesenkov, Astron. zh., Vol. 40, p. 31 (1963).
3. V. G. Fesenkov, Astron. zh., Vol. 40, p. 882 (1963).
4. V. G. Fesenkov, Astron. zh., Vol. 40, p. 1085 (1963).
5. N. B. Divari, Astron. zh., Vol. 43, p. 593 (1966).
6. N. B. Divari and A. S. Asaad, Trudy Astrofiz. inst. Akad. Nauk Kazakh SSR, Vol. 2, p. 52 (1961).
7. N. B. Divari and A. S. Asaad, Astron. zh., Vol. 36, p. 856 (1959).
8. N. B. Divari, Usp. Fiz. Nauk, Vol. 84, p. 75 (1964).
9. E. V. Pyaskovskaya-Fesenkova, Trudy Astrofiz. inst. Akad. Nauk Kazakh SSR, Vol. 3, p. 133 (1962).

SEASONAL VARIATIONS OF THE TELLURIC LINES OF OXYGEN AND WATER VAPOR IN THE SOLAR SPECTRUM

N. I. Kozhevnikov, F. G. Sitnik, and A. T. Khlystov

1. In [1], we examined some of the characteristics of the spectral bands of water vapor in the atmosphere. We used data from observations in 1958 near Moscow (Kuchinsk Astrophysical Observatory) and in the Tien Shan expedition of the P. K. Shternberg State Astronomical Institute near Alma-Ata. These observations showed, among other things, that in mountainous regions the amount of water vapor in the atmosphere varies considerably throughout the day. It was of interest to study the annual variations of the telluric lines. For this, during the summer and winter we made photoelectric recordings of the water-vapor and oxygen spectra on a spectrometer with high dispersion. Some of the results of these studies will be discussed below.

2. When studying telluric-line variations, one must ensure that the observation and recording conditions remain unchanged. Of course, observation of the solar spectrum at one and the same zenith distance would be the best solution of the problem, since this would automatically keep the observation conditions constant and greatly facilitate the task of processing the spectrograms. Unfortunately, weather conditions do not, as a rule, allow this requirement to be satisfied: observations must be made at different solar zenith distances. It becomes necessary to reduce all of the observation data to one air mass.

It is fairly simple to convert a spectrum from air mass M to air mass M*. Let $I_\lambda(M)$ be the intensity of the spectrum at point λ for air mass M. We determine $I_\lambda(M)$ in relative units, taking the intensity in continuum as a unit. Thus, $I_\lambda(M)$ is actually the selective relative transmittance. If k_λ is the selective transmission coefficient for a unit air mass, then the transmission coefficient $k_\lambda(M)$ for air mass M when the Bouguer law holds will be

$$k_\lambda(M) = M k_\lambda. \tag{1}$$

Hence,

$$I_\lambda(M) = e^{-M k_\lambda}. \tag{2}$$

Accordingly, for the other air mass M*

$$I_\lambda(M^*) = e^{-k_\lambda M^*}. \tag{3}$$

From expressions (2) and (3) we obtain

$$I_\lambda(M^*) = e^{-k_\lambda M^* \frac{M^*}{M}} = [I_\lambda(M)]^{M^*/M}. \tag{4}$$

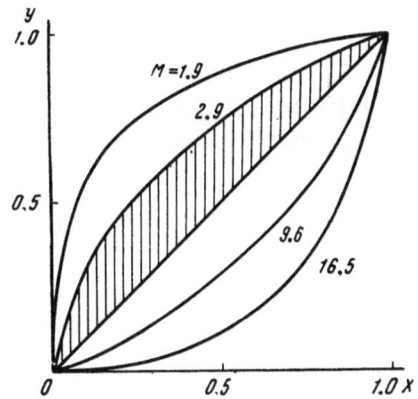

Fig. 1. Nomogram for converting intensity of solar spectrum from one air mass to another.

To check this relation, we used observations with air masses from M = 2 to M = 16. The $I_\lambda(M)$ values were reduced to air mass M* = 4.5. To make the conversion easier, we used the nomogram method for raising the values of $I_\lambda(M)$ to the power M*/M. For this, we plotted a graph of

$$y = x^{M*/M} \qquad (5)$$

for the y range from 0 to 1 and the x range from 0 to 1 (Fig. 1). Figure 1 also shows the relation y = x. We let a unit (on both the x and y axes) be equal to the intensity in continuum on the recording being processed. After plotting a graph as in Fig. 1 (which gives an example of conversion from M = 2.9), the shaded part was cut out and the obtained stencil was superimposed on the spectrogram. Obviously, the ordinates of the spectrogram curve correspond to the line y = x of the stencil, and the result of conversion of an ordinate to M* corresponds to the line $y = x^{M*/M}$. If we move the points of intersection of the spectrogram line with the line y = x vertically up to the line $y = x^{M*/M}$, we find the conversion of $I_\lambda(M)$ to air mass M*.

We studied the telluric lines of the B band of oxygen and the water-vapor lines in the longwave region of the P branch of that band. The results of conversion of the intensity distribution in the two oxygen lines λ 6934.422 and λ 6935.218 from air masses M = 1.9, 2.9, 4.5, 9.6, and 16.5 to air mass M* = 4.5 were consistent within the limits of the standard deviation $\sigma = \pm 0.05$ (in units of the intensity in continuum).

If we know $I_\lambda(M*)$, we can estimate the accuracy with which the Bouguer law holds for the telluric lines and, therefore, the reliability of the results obtained with formula (4). We shall represent the intensity of the telluric line as follows

$$I_\lambda(M) = e^{-k_\lambda M - f_\lambda(M)}. \qquad (6)$$

Here, $f_\lambda(M)$ is a correction to the optical thickness τ_λ, which is not linearly related to M. As is well-known, the Bouguer law can be formulated as

$$\tau_\lambda = k_\lambda M, \qquad (7)$$

where M is the air mass on the ray path, which was calculated from geometric considerations taking into account astronomical refraction, and k_λ is the absorption coefficient for a unit air mass. Deviations from the Bouguer law, therefore, manifest themselves in that the optical thickness becomes a nonlinear function of M.

If we apply relation (4) formally to (6), then

$$I_\lambda(M*)_c = e^{-k_\lambda M* - \frac{M*}{M} f_\lambda(M)}. \qquad (8)$$

At the same time, the observed value $I_\lambda(M*)_o$ will be

$$I_\lambda(M*)_o = e^{-k_\lambda M* - f_\lambda(M*)}. \qquad (9)$$

The value σ gives an estimate of the mean difference between $I_\lambda(M*)_o$ and $I_\lambda(M*)_c$. Processing of the observations has shown that σ is not a function (in the first approximation) of either $I_\lambda(M)$ or M. Therefore, we can write

$$I_\lambda(M*)_c - I_\lambda(M*)_o = \sigma. \qquad (9')$$

Hence,

$$e^{-k_\lambda M^* - \frac{M^*}{M} f_\lambda(M)} - e^{-k_\lambda M^* - f_\lambda(M^*)} = \sigma.$$

If we assume that $f_\lambda(M)$ is small in comparison with unity, divide both sides of the equation by $e^{-k_\lambda M^*}$ and expand the exponential function into a series in powers of the exponent, we obtain

$$f_\lambda(M^*) - \frac{M^*}{M} f_\lambda(M) = e^{k_\lambda M^*} \sigma. \tag{10}$$

Therefore,

$$f_\lambda(M) = \frac{M}{M^*} f_\lambda(M^*) - \sigma \frac{M}{M^*} e^{k_\lambda M^*}. \tag{11}$$

If we factor out M, we have

$$f_\lambda(M) = M \left\{ \frac{f_\lambda(M^*)}{M^*} - \sigma \frac{e^{k_\lambda M^*}}{M^*} \right\}. \tag{12}$$

Thus, the linearity or nonlinearity of $f_\lambda(M)$ with respect to M depends upon the behavior of the difference σ. Since at the accuracy that our observations provided, σ is not a function of M, we can assume that up to M = 16 the Bouguer law holds with an accuracy of ±5% of the value corresponding to τ_λ. It should be remembered that σ includes the recording and measurement errors. We can, therefore, assume that the Bouguer law holds with accuracy better than 0.05.

3. When the solar spectrum is recorded at various times of the year and for different solar zenith distances, the constancy of the instrumental profile of the spectrograph must be checked. For this, we recorded the intensity distribution in the zero order of the spectrum. It was shown in [2] that the intensity distribution in the zero order of the spectrum represents with sufficient accuracy the instrumental function of the spectrograph. After the instrumental profile of the spectrograph has been determined, the line profile must be corrected for the finite width of the instrumental profile. In correcting the line half-width and depth, in the first approximation we can use the conventional formulas that are valid when the line profile and instrumental profile are described by the Gauss curve. This condition is adequately satisfied in the case in question.

Let $\Delta\bar{\lambda}$ and $\Delta\lambda$ be the observed and true line half-widths; $\delta\lambda$ the half-width of the instrumental profile; and $\bar{a}$ and a the observed and true line depths. Then we have

$$(\Delta\bar{\lambda})^2 = \delta\lambda^2 + \Delta\lambda^2, \tag{13}$$

$$a^2 = \bar{a}^2 \left[1 + \left(\frac{\delta\lambda}{\Delta\lambda} \right)^2 \right]. \tag{14}$$

4. Solar spectra with groups of telluric lines were recorded in June-August 1965 and in February 1966 for a number of solar zenith distances. The observations were made on the spectrograph of the solar tower of the P. K. Shternberg State Astronomical Institute. Every time the parameters of the spectrometer (height and width of entrance and exit slits) or the gain and time constant of its recording part, etc., changed the zero order of the spectrum was recorded. Processing of the recordings consisted in correcting the line parameters (half-widths and depths that had been distorted by the instrumental profile of the spectrograph). This correction was made using formulas (13) and (14). Observations of the zero order of the spectrum showed that the half-width of the instrumental profile (in the region of the B band of oxygen) was 0.072, 0.036, and 0.024 Å in the first, second, and third orders, respectively.

Next, the corrected line parameters were reduced to a single air mass. The line depths were reduced to the standard M* by formula (4), and the half-widths were reduced to the half-widths for M* by the formulas in [4].

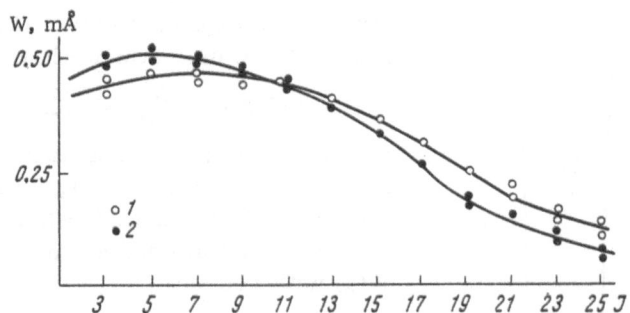

Fig. 2. Equivalent widths of lines of B band of oxygen versus rotational quantum number J: 1) summer; 2) winter.

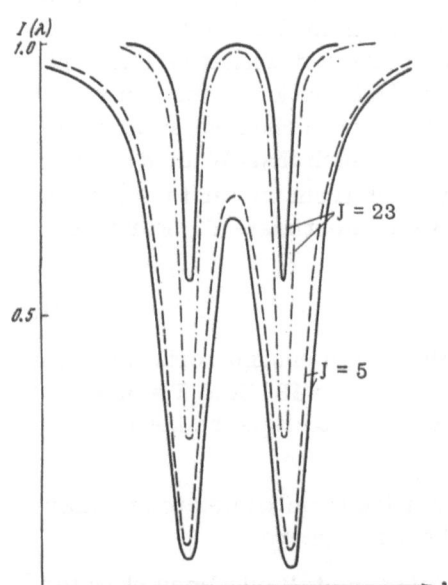

Fig. 3. Seasonal variations of oxygen line profiles (B band). Lines with initial quantum number J = 5 have wavelengths 6888.984 and 6889.903 Å; wavelengths of lines with initial quantum number J = 23 are 6928.727 and 6929.549 Å.

Figure 2 shows the results of observations of the P branch of the B band of O_2. The J values (rotational quantum number) of the lines of the P branch of the B band are plotted on the axis of the abscissas. (See [3-5] on the parameters of the B band.) The equivalent widths W of these lines are plotted on the axis of the ordinates. The W values have been reduced to air mass M = 6. As can be seen from Fig. 2, the equivalent widths of lines with small J increase in winter and decrease in summer. The opposite variation of W is observed for lines with large J.

The central depths and observed half-widths of the lines of the B band of oxygen display a similar variation with the rotational quantum number J in winter and summer.

Figure 3 shows the seasonal variations of the profiles of two line pairs of the P branch of the B band of oxygen. Wavelength is plotted on the axis of the abscissas and line intensity is plotted on the axis of the ordinates (the intensity in continuum is taken as unity). The two lines with J = 5 have wavelengths of 6888.948 and 6889.903 Å, and the lines with J = 23 have 6928.727 and 6929.599 Å.

Thus, the observations show that the telluride lines of oxygen vary throughout the year. The most probable cause of these line variations in the B band is J-state redistribution of the oxygen molecules, due to the difference between the mean temperatures in winter and summer. For the rotational-level distribution we have the following formula [5]:

$$N(J) = N \frac{hcB}{kT} (2J + 1) e^{-BJ(J+1)hc/kT}. \tag{15}$$

The values N(J) and N (i.e., the number of molecules in the J state and the total number of molecules) cannot be measured directly. Therefore, let us convert (see [6]) to the equivalent widths of the intense lines with the formula

$$N(J) = f(J) [W(J)]^2, \tag{16}$$

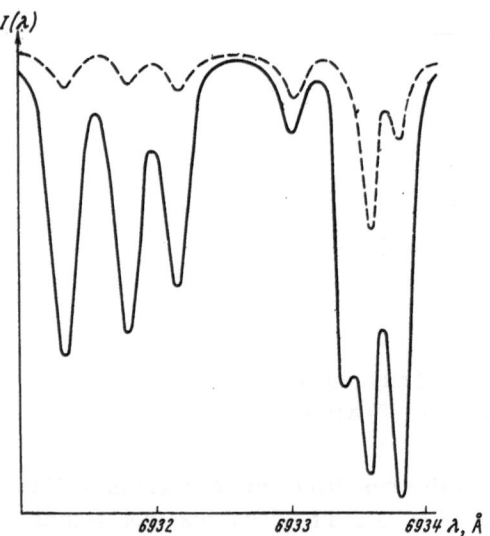

$I(\lambda)$

6932   6933   6934 λ, Å

Fig. 4. Seasonal variations of water-vapor line profiles in region 6931-6934 Å.

where $f(J)$ is a function of the parameters of the J state. The explicit form of the function $f(J)$ is not important to us. Now we have

$$[W(J)]^2 = \frac{N}{f(J)} \cdot \frac{hcB}{kT} (2J+1) e^{-BJ(J+1)hc/kT}, \quad (17)$$

where h is Planck's constant; c the velocity of light; k the Boltzmann constant; T the absolute temperature; and B a constant, which is equal to 1.44 cm^{-1} for an oxygen molecule. We find the derivative of W(J) with respect to T,

$$\frac{\partial W(J)}{\partial T} = \sqrt{\frac{N(2J+1)\alpha}{f(J)}} \, e^{-\frac{\alpha J(J+1)}{2T}} \frac{1}{2T^{3/2}} \left\{ \frac{\alpha(J+1)J}{T} - 1 \right\}, \quad (18)$$

where $\alpha = hcB/k$.

It follows from formula (18) that, as T increases, the derivative $\partial W(J)/\partial T$ is negative for small J and positive for large J. The observation data, therefore, are in indirect agreement with theory. We can also make a quantitative check of our hypothesis that redistribution of the equivalent widths in the B band is due to changes in air temperature. From formula (18) it follows that for every T there is a J_0 such that $\partial W(J)/\partial T = 0$, i.e., the equivalent line width with this J_0 must not change with transition from winter to summer. If we know the quantum number J_0 for which W(J) is constant, we can estimate T. For this, we let $\partial W(J)/\partial T$ equal zero and, hence, we find

$$T = \alpha J_0 (J_0 + 1). \quad (19)$$

For an oxygen molecule, $\alpha = 2.07$. The value J_0 is not determined with enough certainty from Fig. 3 from $J_0 = 9$ to $J_0 = 13$. Accordingly, for T we have the limits T = 180°K to T = 360°K. On the average, we obtain 270°K. This T value is of the same order of magnitude as the mean temperature of the earth's atmosphere.

Thus, the line variation in the B band of oxygen with transition from winter to summer should be attributed to variation of the mean temperature of the atmosphere.

5. Figure 4 shows a region of the solar spectrum containing the telluric lines of water vapor. The solid line corresponds to summer observations and the broken line to winter observations. Both the winter and summer data have been reduced to air mass M = 4.5. It can be seen that the variation in the amount of atmospheric water vapor is very considerable.

In conclusion, we must say a few words about the place held by seasonal transmittance variations in the overall picture of the optical instability of the atmosphere. Variations in atmospheric transmittance (and processes leading to these variations) evidently form a distinct hierarchy of time and space structures. Star scintillation and vibration of the solar-limb image have a period of 10^{-2}-10^1 sec and owe their origin to local atmospheric heterogeneities on the order of 10-10^3 cm [7, 8]. The daily variations in transmittance are probably due to interchange of air masses as well as variation of the thermal conditions and the development of convective instability in the lower layers of the atmosphere [9-11]. These processes have spatial scales of from 10^2 m to 10^2 km. Then come the seasonal variations in both continuous and selective transmittance. They are due to the annual variation of thermal conditions and to change of the prevailing type of air mass [10]. These variations have a time scale on the order of 3-4 months and a spatial scale on the order of hundreds and thousands of km. Finally, there are transmittance variations with a period on the order of 11 years, which are related to the solar cycle and are global in nature [12].

LITERATURE CITED

1. E. A. Makarova, G. F. Sitnik, and N. I. Kozhevnikov, Soobshcheniya GAISh, No. 126 (1963).
2. N. I. Kozhevnikov and E. F. Rizov, Soobshcheniya GAISh, No. 126 (1963).
3. A. S. Tibilov and V. K. Prokof'ev, Izv. KrAO, Vol. 32, p. 3 (1966).
4. N. I. Kozhevnikov, G. F. Sitnik, M. A. Klyakotko, and G. A. Porfir'eva, Astronomicheskii vestnik (in press).
5. G. Herzberg, Molecular Spectra and Molecular Structure, in: Spectra of Diatomic Molecules, Vol. 1, 2nd ed., Van Nostrand, Princeton, N. J. (1950).
6. A. Unsöld, Physics of Stellar Atmospheres [Russian translation], IL (1949).
7. N. I. Kozhevnikov, Nauchnye doklady vysshei shkoly, No. 3 (1958).
8. V. A. Krat, Transactions of Conference on Star Scintillation [in Russian], Izd. Akad. Nauk SSSR (1959).
9. G. F. Sitnik, Astron. zh., Vol. 42, No. 5, p. 996 (1965).
10. G. F. Sitnik and R. N. Khmeleva, Soobshcheniya GAISh, No. 109 (1960).
11. E. V. Pyaskovskaya-Fesenkova, Investigation of Light Scattering in the Earth's Atmosphere [in Russian], Izd. Akad. Nauk SSSR (1957).
12. N. I. Kozhevnikov, A. B. Delone, and G. F. Sitnik, Solnechnye dannye, No. 3 (1965).

METHOD FOR DETERMINING
ATMOSPHERIC EXTINCTION

A. P. Sarychev

The compilation of a fundamental catalogue of star magnitudes with the aid of a modern photoelectric photometer requires careful determination of light absorption in the atmosphere. The anticipated error of photoelectric measurement of brightness is $0^{m}005$. Such high accuracy compels one to take into account the variation in transmittance throughout the night.

In general form, the extinction coefficient α is a function of the observation time t and the extra-atmospheric color index C_0

$$\alpha = \alpha(t, C_0).$$

Assuming that this function is linear relative to C_0, it is usually converted to a single color index $\tilde{C}$

$$\alpha(t, \tilde{C}) = \alpha(t, C_0) - \gamma(C_0 - \tilde{C}),\tag{1}$$

where $\gamma = \partial\alpha/\partial C_0$ is the gradient of the extinction coefficient as a function of the color index. For a known γ, expression (1) makes it possible to compare the extinction coefficients found for stars of different spectral classes.

Now let us make certain simplifying assumptions about the variation of α with time. We shall assume that the instantaneous transmission coefficient is the same over the entire sky; this frees us from local heterogeneities. The second assumption utilizes the photometrically detected smoothness of the curve $\alpha(t)$. We mentally break up this line into pieces. A straight line gives a good approximation of a small segment of a smooth curve. We shall assume that α varies linearly within any short time segment. Let a large number of brightness measurements be made throughout the night. If brightness determinations are made sufficiently frequently, then the extinction coefficient given by formula (1) of the i-th observation can be considered a result of linear interpolation between its values in the preceding (i − 1) and succeeding (i + 1) observations

$$\alpha(t_i, \tilde{C}) = \alpha(t_{i-1}, \tilde{C}) + \frac{\alpha(t_{i+1}, \tilde{C}) - \alpha(t_{i-1}, \tilde{C})}{t_{i+1} - t_{i-1}}(t_i - t_{i-1}).\tag{2}$$

Application of formula (2) to the aggregate of measurements is valid when the error due to nonlinearity of $\alpha(t)$ over twice the average observational interval $\Delta t \approx t_{i+1} - t_{i-1}$ is much smaller than the measurement error. In practice, (2) is invalidated only by random errors. Linear interpolation in the above statistical sense can be achieved by limiting the maximum interval between observations, since the measurement error is constant and the error due to nonlinearity can be reduced by reducing Δt. A measurement interval of 15–20 min is normally adequate.

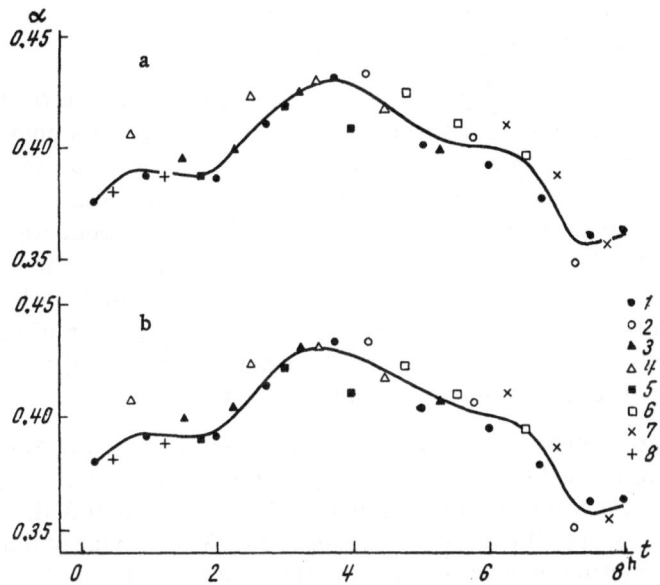

Fig. 1. Variation of extinction coefficient α (reduced to $B - V = +0.2$) during night of 19-20 September 1965: (a) by proposed method; (b) by V. B. Nikonov's method; 1) α Cyg; 2) α Aur; 3) α Per; 4) α Lyr; 5) α Aql; 6) β Tau; 7) γ Ori; 8) η U Ma.

On the basis of (1) and the Bouguer formula, we can express $\alpha(t_i, \widetilde{C})$ in terms of the observed and unknown values

$$\alpha(t_i, \widetilde{C}) = \frac{m_i - m_{0i}}{F_i(z)} - \gamma(C_{0i} - \widetilde{C}),\tag{3}$$

where m is the measured brightness and m_0 is the extra-atmospheric brightness.

Substitution of (3) into (2) gives

$$\frac{t_{i+1} - t_{i-1}}{F_i(z)} m_{0i} - \frac{t_i - t_{i-1}}{F_{i+1}(z)} m_{0i+1} - \frac{t_{i+1} - t_i}{F_{i-1}(z)} m_{0i-1} +$$

$$+ [(t_{i+1} - t_{i-1}) C_{0i} \gamma - (t_i - t_{i-1}) C_{0i+1} \gamma - (t_{i+1} - t_i) C_{0i-1} \gamma -$$

$$- \left(\frac{t_{i+1} - t_{i-1}}{F_i(z)} m_i - \frac{t_i - t_{i-1}}{F_{i+1}(z)} m_{i+1} - \frac{t_{i+1} - t_i}{F_{i-1}(z)} m_{i-1} \right)] = 0,\tag{4}$$

where m_{0i} is the extra-atmospheric brightness in the i-th observation. Note that the subscript i indicates the observation number, but it is completely interchangeable with the star number. In processing the data, the specific number of the star replaces i. Equation (4) contains the extra-atmospheric magnitudes of three stars; the coefficients of m_0 and the free term in the brackets are calculated. This equation is valid for every i except the first and the last, where interpolation is impossible. After calculation of the parameters of (4) corresponding to i = 2, 3, 4,..., we have the system of linear equations

$$a_{11} m_{01} + a_{12} m_{02} + \ldots + a_{1k} m_{0k} + b_1 = 0,$$
$$a_{21} m_{01} + a_{22} m_{02} + \ldots + a_{2k} m_{0k} + b_2 = 0,$$
$$\cdot \quad \cdot \quad \cdot \quad \cdot \quad \cdot \quad \cdot \quad \cdot \quad \cdot \quad \cdot \quad \cdot \quad \cdot \quad \cdot \quad \cdot \quad \cdot \quad \cdot \quad \cdot \quad \cdot \quad \cdot \tag{5}$$
$$a_{n1} m_{01} + a_{n2} m_{02} + \ldots + a_{nk} m_{0k} + b_n = 0,$$

where m_{01}, m_{02},...,m_{0k} are the unknown brightnesses of the k observed stars; a and b are calculated from (4), and a = 0 for stars that do not enter into (4) (obviously, in any equation of (5) only three coefficients a can differ from zero).

It is permissible to combine interpolation Eqs. (5) in a system, since they are successively related to one another by common unknowns; the next equation always contains two stars of the preceding one. The number of equations n in the system is equal to the number of brightness measurements made on a night minus two. If observations of a group of stars are made over several nights and the systems of equations for these nights are connected in pairs by common stars into a continuous chain, then system (5) must contain all of the obtained interpolation equations and n will be equal to the total number of measurements minus twice the number of nights. When the same stars recur in observations, the number of unknowns k (the number of stars in the program) will be less than n. Solution of system (5) by the method of least squares gives the reduced, extra-atmospheric star magnitudes.

System (5) represents an intermediate step in the joint solution of the set of Eqs. (2) and (3); it is formed by eliminating the unknowns α from (2) with the aid of (3). It is easy to show the linear independence of any Eq. (2) from the others, and also the mutual linear independence of Eqs. (2) and (3). Therefore, the absence among Eqs. (3) of linearly dependent values is sufficient to show that no equation of system (5) is a linear combination of the others. The independence of each equation in (3) must be ensured by prearranging the observation conditions: repeated measurements of the brightness of a star must be made with substantially different air masses. In setting up a program of observations, one must remember the requirement that the equations for individual nights be related.

After the unknowns m_0 have been determined, a graph of the extinction coefficient as a function of time should be plotted and used to check the validity of the initial assumptions. The $\alpha(t_i)$ values from formula (3) are plotted and a continuous curve is drawn through them (Fig. 1). From the graph, it is easy to judge whether the spread of the points exceeds the deviation of the curve from linearity, or whether interpolation formula (2) is statistically valid. The graph also reflects the effect of local transmittance heterogeneities, which, together with the measurement error, cause the points to deviate from the median curve. Constant jumps of the points of any star up (or down) indicate that the star was observed only in regions whose transmittance was lower (or higher) than the mean. In this case, reduction involves a systematic error. Usually, manifestations of local extinction are equalized by observations in parts of the sky with opposite deviations of transmittance.

In order to calculate the free term b in (5), we must know the color indices C_{01}, C_{02},...,C_{0k} or make special measurements, in particular, by a method similar to the one being considered here. Just as before, we subject the reduced (to $\widetilde{C}$) extinction coefficient of the color index $\alpha^*(t_i, \widetilde{C})$ to the interpolation formula

$$\alpha^*(t_i, \widetilde{C}) = \alpha^*(t_{i-1}, \widetilde{C}) + \frac{\alpha^*(t_{i+1}, \widetilde{C}) - \alpha^*(t_{i-1}, \widetilde{C})}{t_{i+1} - t_{i-1}}(t_i - t_{i-1}). \tag{6}$$

The Bouguer law and the expansion of (1) with the new gradient $\gamma^* = \partial\alpha^*(t, C_0)/\partial C_0$ remain valid

$$\alpha^*(t_i, \widetilde{C}) = \frac{C_i - C_{0i}}{F_{i(z)}} - \gamma^*(C_{0i} - \widetilde{C}); \tag{7}$$

TABLE 1

Star	Number of observations	m_0 in photometer system	
		Nikonov's method (75 equations)	Proposed method (100 equations)
α Cyg	35	$-0^{m}.705$	$-0^{m}.701 \pm 0.008$
α Aur	14	-1.217 ± 0.006	-1.215 ± 0.011
α Per	13	$+0.205 \pm 0.009$	$+0.214 \pm 0.014$
α Lyr	12	-2.023 ± 0.004	-2.022 ± 0.012
α Aql	11	-1.073 ± 0.006	-1.069 ± 0.013
β Tau	10	-0.523 ± 0.012	-0.526 ± 0.013
γ Ori	7	-0.631 ± 0.010	-0.633 ± 0.015
η U Ma	5	-0.337 ± 0.007	-0.335 ± 0.021
α Boo	1	-1.406	-1.384 ± 0.050

where C is the measured color index and C_0 is the extra-atmospheric color index. Substitution of (7) into (6) gives

$$\left[\left(t_{i+1} - t_{i-1} \right) \left(\frac{1}{F_i} + \gamma^* \right) \right] C_{0i} - \left[\left(t_i - t_{i-1} \right) \left(\frac{1}{F_{i+1}} + \gamma^* \right) \right] C_{0i+1} -$$

$$- \left[\left(t_{i+1} - t_i \right) \left(\frac{1}{F_{i-1}} + \gamma^* \right) \right] C_{0i-1} - \left[\frac{t_{i+1} - t_{i-1}}{F_i} C_i - \frac{t_i - t_{i-1}}{F_{i+1}} C_{i+1} - \frac{t_{i+1} - t_i}{F_{i-1}} C_{i-1} \right] = 0. \quad (8)$$

Equation (8) contains the unknown extra-atmospheric color indices of three stars C_{0i-1}, C_{0i}, and C_{0i+1}. All of the values in the brackets are found directly from measurements, with the exception of the small correction γ^*, which is calculated independently. Formula (8) is similar in structure to formula (4), and the results of analysis of the method for reducing star magnitudes are automatically carried over to the case of the color index.

The proposed reduction method was checked on a series of observations made at the Crimean Astrophysical Observatory on 15-16, 17-18, 18-19, and 19-20 September 1965 with the aim of compiling a fundamental catalogue of a number of bright stars. The brightness measurements were made with a photoelectric photometer relative to a reference radioluminescent source. The color system of the photometer was similar to the B system of Johnson and Morgan; λ_{eff} was about 4400 Å for isoenergetic radiation. The observation program was selected on the basis of the fact that we intended to determine extinction by V. B. Nikonov's method [1-3]. The obtained material was processed by the proposed method and by Nikonov's method as modified by A. S. Sharov and V. P. Arkhipova [4]. The working value of the gradient $\gamma = -0.03$ was selected on the basis of the mean spectral transmittance of the atmosphere. The results are summarized in the table. The two methods gave practically the same extra-atmospheric star magnitudes. The difference in the mean-square errors is due to the different principles according to which they were determined: convergence of the equations of an individual star in Nikonov's method [4] and general convergence of all equations [5] in the proposed method. The real error of reduction, as the spread of points in the figure shows, is approximately the same, i.e., the effectiveness of the methods is very similar.

Our method involves more time-consuming calculations than does Nikonov's method. But this drawback can be overcome by using electronic computers.

The advantage of our method is that it does not use a standard star. In Nikonov's method, the standard star must be observed more often than the others, and its brightness enters each equation jointly with the brightness of any star. Such artificial weighting of any star is dangerous, since it increases the distorting effect of local turbidity on the reduction of the

A. P. SARYCHEV

other stars. While observations of the standard star in Nikonov's method do not form characteristic equations, the proposed method increases their number in the system (see Table 1), and, therefore, makes more economical use of the observations. Unfortunately, these advantages were only partially realized in the example given here, which was based on Nikonov's program.

I thank A. S. Sharov for very valuable advice, and also N. I. Ushakova for assistance in the observations and processing.

LITERATURE CITED

1. V. B. Nikonov, Doklady Akad. Nauk SSSR, Vol. 45, p. 151 (1944).
2. V. B. Nikonov, Byull. Abastum. obs., No. 14 (1953).
3. V. B. Nikonov and E. K. Nikonova, Izv. KrAO, Vol. 9, p. 135 (1952).
4. A. S. Sharov and V. P. Arkhipova, Soobshcheniya GAISh, No. 130, p. 3 (1952).
5. B. M. Shchigolev, Mathematical Processing of Observations [in Russian], Moscow (1962), p. 239.

DETERMINING THE TRANSMISSION COEFFICIENT AND DEGREE OF OPTICAL STABILITY OF THE EARTH'S ATMOSPHERE

E. V. Pyaskovskaya-Fesenkova

As is well known, the Bouguer long method gives the correct value of the transmission coefficient only when atmospheric transmittance remains constant during observations. Therefore, we must have a way to monitor the optical stability of the atmosphere. The Bouguer line does not provide such a possibility. The relative aureole, i.e., the ratio of the illuminations from the solar aureole and from the sun, can serve as a criterion of optical stability. In 1933, V. G. Fesenkov showed theoretically that the relative aureole is linearly related to the atmospheric mass in the direction of the sun, and, therefore, when it is determined at a number of different solar zenith distances, it gives a straight line that passes through the coordinate origin. Any even slight change in the optical properties of the atmosphere will cause individual determinations to depart from this line [1]. In this respect, the line of the relative solar aureole is a more sensitive characteristic of these properties than is the Bouguer line.

Later, this problem was studied in detail by the author, and refined and added to, particularly with regard to the optical instability of the atmosphere. On optically unstable days, the line of the relative aureole passes above or below the coordinate origin, depending upon the direction in which the optical properties change. On an optically stable day, the relative aureole reduced to a unit atmospheric mass gives a straight line parallel to the axis of the abscissas, but on an unstable day, this line is disrupted. This line is a very sensitive criterion of optical stability [2, 3].

It must be added that a one-to-one relationship does not exist between the relative aureole and the transmission coefficient. As our observations have shown, the one-to-one relationship between these values can be destroyed when the nature of the aerosols (dry or wet) changes. This can happen, for example, when polar air is replaced by tropical, and vice versa. Thus, the relative aureole can differ for the same transmittance. There are only very rare occasions, however, when atmospheric transmittance remains constant when the nature of the aerosols changes during one day. Therefore, the relative aureole can serve as a criterion of transmittance stability.

The aureole photometer designed by V. G. Fesenkov, to which the author added photocell-temperature recording, makes it possible to determine the transmission coefficient and to monitor its stability, since this photometer can measure separately the fluxes from the sun and from the solar corona. This instrument is simple in design. It contains no optics, with the exception of the filters, which is a great advantage, and its provides rather good results,

151

when the instrument itself and the photocell have been carefully studied and the galvanometer has been properly selected. Wideband filters are used, and the detector is a selenium photocell. Interference filters and a photomultiplier may be substituted, however.

As is known, when measuring solar radiation, not only direct solar radiation but also radiation scattered by the aureole enters the detector, and this introduces an error into the transmission coefficient. But since V. G. Fesenkov's aureole photometer can measure both solar radiation and the aureole, the author has attempted to determine the error in the transmission coefficient p caused by the aureole. The circular zone cut out by the photometer in aureole observations has angular radii of 3°4' and 1°58' without penumbra and 3°36' and 0°52' with penumbra. If we know the brightness distribution in the aureole and the galvanometer reading when it is measured in this zone, we can find the brightness of the entire aureole in the same units. Then we subtract this brightness from the galvanometer reading for solar radiation. This gives the radiation without the effect of the aureole, and by plotting the Bouguer line, we obtain the true transmission coefficient.

The brightness distribution in the aureole was found by the Van de Hulst formula, which can be used when $0°5 < \vartheta < 5°$,

$$B(\vartheta) = A\vartheta^{-q}, \tag{1}$$

where $B(\vartheta)$ is the brightness of the aureole at angular distance ϑ from the sun; A is a constant; and q varies, according to Van de Hulst, from 1.2 to 1.8. V. E. Pavlov [4] and T. P. Toropova [5] have studied this formula and have found it to be in good agreement with observations. According to Pavlov, q varies from 0.96 with a blue filter ($\lambda_{eff} = 447$ mμ) on a clear day to 1.92 with a red filter ($\lambda_{eff} = 630$ mμ) on days with high turbidity.

We determined the value

$$K = \frac{2\pi \int_{0°5}^{3°} B(\vartheta) \sin \vartheta d\vartheta}{2\pi \int_{2°}^{3°} B(\vartheta) \sin \vartheta d\vartheta} = \frac{\int_{0°5}^{3°} \vartheta^{1-q} d\vartheta}{\int_{2°}^{3°} \vartheta^{1-q} d\vartheta}, \tag{2}$$

where $B(\vartheta)$ is given by expression (1). If we multiply the galvanometer reading for aureole measurement by K, we obtain the brightness of the aureole in the same units within the limits of ϑ from 3 to $0°5$ for the corresponding q. Values of K for a number of q values are given below:

q	0.96	1.00	1.20	1.40	1.60	1.80	1.92
K	2.46	2.50	2.75	3.05	3.41	3.87	4.18

Then we calculated the solar radiation as measured with blue ($\lambda_{eff} = 445$ mμ) and red filters ($\lambda_{eff} = 636$ mμ) and determined the transmission coefficients for 18 September and 5 and 7 October 1949 taking into account the aureole. All three days were clear, with large bright, golden aureoles. Since we did not take into account the aureole at $\vartheta < 0°5$, the overestimated value q = 1.92 was used for the blue and red rays (Table 1).

Table 2 shows the transmission coefficients as determined from direct observations with (p') and without (p") the effect of the aureole, and also $\Delta p = (p' - p")/p"$.

The tables give the percent errors $\Delta F_\odot$ as the measured solar radiation for various atmospheric masses toward the sun $m_\odot$, and also the errors Δp in the transmission coefficients p

TABLE 1

18 September, P.M.			5 October, P.M.			7 October, P.M.		
$z_\odot$	$m_\odot$	$\Delta F_\odot$, %	$z_\odot$	$m_\odot$	$\Delta F_\odot$, %	$z_\odot$	$m_\odot$	$\Delta F_\odot$, %
Blue filter			Blue filter			Blue filter		
76°10′	4.12	4.3	82°24′	7.20	5.6	70°12′	2.93	2.3
81 07	6.25	6.7	85 02	10.44	8.0	77 54	4.68	3.7
84 00	8.91	10.1	85 30	11.32	8.8	81 18	6.36	5.1
85 31	11.38	13.9	86 02	12.53	9.6	84 00	8.89	7.3
86 19	13.23	17.3	86 38	14.18	11.2	84 36	9.75	8.0
			Red filter			Red filter		
			82°12′	7.07	4.9	60°15′	2.01	1.2
			84 13	9.18	6.2	76 18	4.17	2.7
			85 53	12.18	8.3	81 00	6.18	4.1
			86 28	13.68	9.1	83 48	8.65	5.8
			87 19	16.53	11.1	84 29	9.58	6.4

Note. $\Delta F_\odot = (F'_\odot - F''_\odot)/F''_\odot$ ($F'_\odot$ and $F''_\odot$ are the observed solar radiation and the solar radiation taking into account the aureole, respectively).

as determined by the Bouguer long method, which were introduced by scattered light from the aureole. As can be seen from these tables, the effect of the aureole on the transmission coefficient as determined by the Bouguer long method is negligible, since under normal atmospheric conditions it is within the measurement error.

Observations have been made since 1943 with this instrument by several researchers in the Department of Atmospheric Optics of the Astrophysical Institute of the Academy of Sciences of the Kazakh SSR. As an example, we shall give the results of some observations on optically stable and unstable days. Figures 1 and 2 show results of afternoon observations by T. P. Toropova at the mountain observatory of the Astrophysical Institute on 7 and 5 October 1949 with blue, green, and red filters (the effective wavelengths of the photocell-filter system were 445, 546, and 636 mμ, respectively). The Bouguer lines are shown in Figs. 1a and 2a. The points fit them very well, but this does not mean that the atmosphere was optically stable on those days. To judge this, we must turn to the relative-aureole lines, which are shown in Figs. 1b and 2b. In the former, the line passes through the coordinate origin, and this indicates an optically stable atmosphere. In the latter, the line intersects the axis of the ordinates above the coordinate origin, which means that on the afternoon of that day the relative aureole decreased gradually as $m_\odot$ increased. The points are lower than they would have been if stability had been preserved, and transmittance increased gradually. If the relative aureole had increased gradually as $m_\odot$ increased, the points would have been higher and the line would have intersected the axis of the ordinates below the origin. Transmittance would have decreased gradually. Thus, a glance at the relative-aureole line shows immediately in which direction the optical properties changed or whether the atmosphere remained stable within the error limits. The ordinate of the point on the relative-aureole line at $m_\odot = 0$ can serve as a measure of that instability. The more rapid the change in optical properties, the greater the absolute value of this ordinate. If this ordinate is positive, the transmittance is greater at higher $m_\odot$ than at lower. If the ordinate is negative, the transmittance is less at higher $m_\odot$ than at lower. This is true for both morning and afternoon observations [3].

It should be pointed out that the reflections in the photometer telescope must be studied. In our case, reflections were negligible, so passage of the relative-aureole line through the coordinate origin indicates an optically stable day. If reflections are not negligible, on an optically stable day the line will pass above the coordinate origin, and the ordinate at $m_\odot = 0$

TABLE 2

Date	Filter	p'	p''	Δp, %
18 September	blue	0.739	0.731	1.1
5 October	blue	0.767	0.763	0.5
5 October	red	0.870	0.864	0.7
7 October	blue	0.783	0.776	0.9
7 October	red	0.881	0.872	1.0

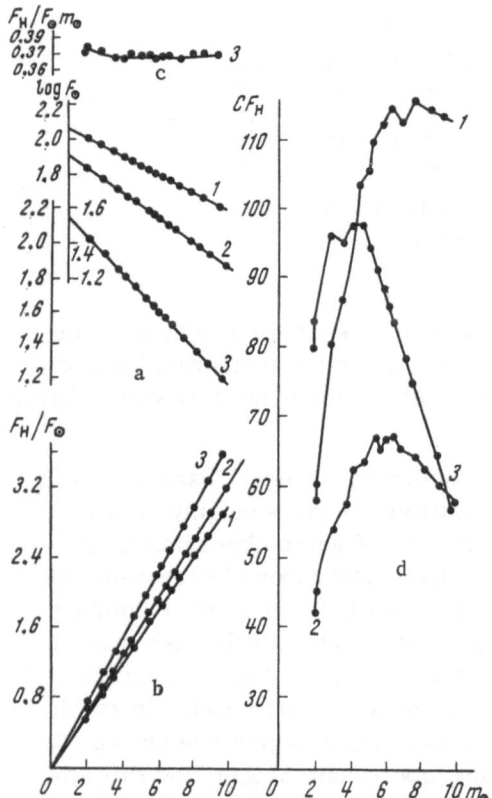

Fig. 1. Afternoon of 7 October 1949:
1, 2, 3) red, green, and blue filters.
a) Bouguer lines, $p_r = 0.881$, $p_g = 0.843$, $p_b = 0.783$; b) relative-aureole lines; c) relative aureole for unit $m_\odot$; d) values proportional to absolute aureole, $p_r = 0.881$, $p_g = 0.846$, $p_b = 0.783$.

will be the constant of the instrument, which must be known and taken into account.

The relative aureole reduced to a unit atmospheric mass $F_H/F_\odot m_\odot$ provides a more detailed indication of optical stability. On a stable day, this value plotted as a function of $m_\odot$ gives a straight line parallel to the axis of the abscissas, as can be seen from Fig. 1c, for 7 October. On this day, the variation of $F_H/F_\odot m_\odot$ did not exceed 7%.

A day when $F_H/F_\odot m_\odot = \mu(\vartheta)$ does not vary by more than 10% we consider a stable day with regard to transmittance. This is determined by the fact that a 10% variation of the directional-scattering factors $\mu(\vartheta)$ for small (aureole) scattering angles changes the transmission coefficient by only tenths of a percent.

On 5 October, the variation of $F_H/F_\odot m_\odot$ during the observations reached 17% — the day was optically unstable. But in the interval $7.20 < m_\odot < 14.18$, this variation did not exceed 4%.

Thus, $F_H/F_\odot m_\odot$ as a function of $m_\odot$ allows us to isolate time intervals with optical stability, and this has practical value.

Observations of the solar aureole also allow us to determine the atmospheric transmission coefficient. The brightness of the aureole increases from morning to noon, reaches a maximum at some atmospheric mass $m'_\odot$, and then decreases. After noon, the brightness of the aureole again increases, again reaches a maximum at, as a rule, a different atmospheric mass, and then decreases. Values proportional to the brightness of the aureole (CF_H), where F_H is the radiation flux from the aureole and C is the proportionality factor, which varies according to the filter, are plotted on the axis of the ordinates in Figs. 1d and 2d.

The brightness of the sky $B(\vartheta)$ on the solar almucantar and, therefore, of the aureole, is given by the well-known formula

$$B(\vartheta) = E_\odot^0 p^{m_\odot} m_\odot \mu(\vartheta),\qquad(3)$$

where $E_\odot^0$ is the solar constant; p the atmospheric transmission coefficient; and $\mu(\vartheta)$ the directional-scattering factor.

If we differentiate this formula with respect to $m_\odot$ and let the result equal zero, we find $m_\odot = m'_\odot$, at which the brightness of the aureole reaches its maximum. On an optically stable day, p and $\mu(\vartheta) = $ const. As a result, we obtain

$$\ln p = -\frac{1}{m'_\odot}.\qquad(4)$$

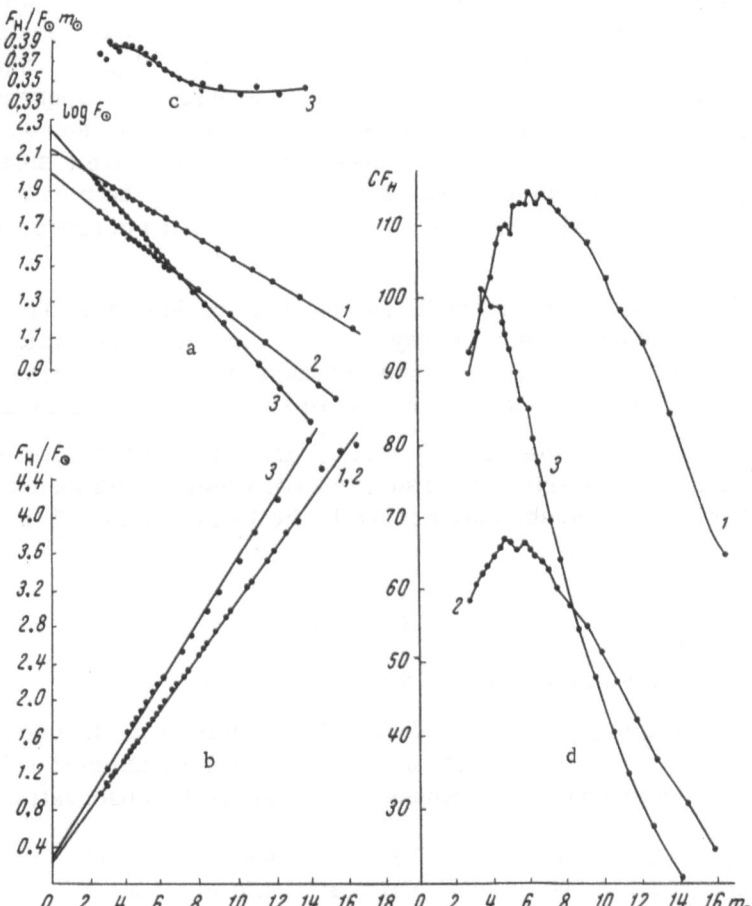

Fig. 2. Afternoon of 5 October 1949: 1, 2, 3) red, green, and blue filters. a) Bouguer lines, $p_r = 0.870$, $p_g = 0.832$, $p_b = 0.767$; b) relative-aureole lines; c) relative aureole for unit $m_\odot$; d) values proportional to absolute aureole, $p_r = 0.857$, $p_g = 0.825$, $p_b = 0.739$.

From this expression it follows that, the smaller $m'_\odot$, the smaller p [3, 6]. The transmission coefficients as determined by formula (4) and the Bouguer method agree within the error limits, as can be seen from Fig. 1. On optically unstable days, $\mu(\vartheta)$ and p vary with $m_\odot$. In this case, the relationship between p and $m'_\odot$ is more complex:

$$\ln p = -\left[\frac{1}{m'_\odot} + m'_\odot \frac{dp}{pdm} + \frac{d\mu}{\mu dn_i}\right],\tag{5}$$

where dp/pdm and $d\mu/\mu dm$ have opposite signs, which more or less balances the effects of the second and third terms of (5).

We shall measure the solar radiation and the solar aureole simultaneously at various $z_\odot$ on an optically unstable day when transmittance changes gradually and then determine p by the Bouguer long method and by formula (4), therefore ignoring the last two terms of formula (5). In this case, the Bouguer long method will give a fictitious p, which, as will be seen below, can with a small error be related to a time with high $z_\odot$. At other $z_\odot$ values, this p is either too large or too small, depending upon the direction in which p changes. At the same time, the p determined by formula (4) is always smaller in the first case and larger in the second case than that determined by the Bouguer method. Thus, on an optically unstable day, the p given by formula (4) corresponds to a time with $m_\odot \neq m'_\odot$. Figure 2 shows that, on the day in question, the transmittance increased from noon to evening. Consequently, the Bouguer method gave an exaggerated p, which can be related to a time when the sun was close to the horizon. Formula (4) (the method of maximum aureole brightness) gave a smaller p, which corresponds to an $m_\odot$ that is close to $m'_\odot$. The difference between the p values was only 1.5% (red filter), 0.9% (green filter), and 3.7% (blue filter) relative to the corresponding mean value of p.

It must be noted, however, that the brightness of the aureole plotted as a function of the atmospheric mass toward the sun does not always have a maximum. On optically unstable days when there is a considerable gradual change in transmittance, this curve can increase monotonically with $m_\odot$ or give fictitious maxima due to sudden variations in transmittance.

On an optically unstable day, the Bouguer method gives not only a fictitious p but also a fictitious value of extra-atmospheric solar radiation $F_\odot^0$, which is obtained by extrapolation of the Bouguer line to $m_\odot = 0$. The absolute errors in the logarithms of $F_\odot^0$ and p ($\varepsilon_{\log F_\odot^0}$ and $\varepsilon_{\log p}$) are related as

$$\varepsilon_{\log p} = \pm \frac{1}{m_\odot} \varepsilon_{\log F_\odot^0},\qquad\qquad\qquad\qquad (6)$$

which was derived on the basis of the Bouguer formula.

From graphs with Bouguer lines obtained over 3.5 months at the mountain observatory of the Astrophysical Institute, we determined for various $m_\odot$ the maximum relative errors of $p(\delta_p)$ due to a gradual linear change in transmittance during the observations:

$m_\odot$	1.5	2	3	5	5,6	10	15
$z_\odot$	48°.3	60°.0	70°.7	78°.7	80°.0	84°46'	86°54'
δ_p, %	4.6	3.5	2.3	1.4	1.2	0.7	0.5

The error of p is a function of the time to which the transmission coefficient determined by the Bouguer method is coordinated: δ_p decreases as $m_\odot$ (or $z_\odot$) increases [3].

Examination of the above data shows that the Bouguer long method gives a sufficiently accurate p during the morning and evening hours at high $z_\odot$ when there is a gradual linear variation of transmittance. At the mountain observatory of the Astrophysical Institute, even when $m_\odot = 3$ ($z_\odot = 70°.7$) the error in the transmission coefficient as determined by the Bouguer method on an optically unstable day is, as a rule, on the order of 2%. This applies to "normal" days, i.e., days without thick haze, fog, etc.

LITERATURE CITED

1.   V. G. Fesenkov, Astron. zh., Vol. 10, No. 3 (1933).
2.   E. V. Pyaskovskaya-Fesenkova, Izv. Akad. Nauk Kazakh SSR, seriya astron. i fiz., No. 5 (1951).
3.   E. V. Pyaskovskaya-Fesenkova, Investigation of Light Scattering in the Earth's Atmosphere [in Russian], Izd. Akad. Nauk SSSR (1957).
4.   V. E. Pavlov, Astron. zh., Vol. 41, No. 1 (1964).
5.   T. P. Toropova, Trudy Astrofiz. inst. Akad. Nauk Kazakh SSR, Vol. 3 (1962).
6.   E. V. Pyaskovskaya-Fesenkova, Astron. zh., Vol. 22, No. 6 (1945).

PHOTOELECTRIC APPARATUS FOR STUDYING
ATMOSPHERIC LIGHT SCATTERING

G. S. Isaev

Investigation of the optical properties of the atmosphere under real conditions requires the greatest possible amount of experimental data obtained in the shortest possible time period. In this connection, the need for automatic high-speed photoelectric apparatus that would allow us as much information as possible about the optical state of the atmosphere to be obtained has often been expressed [1]. Two years of work at the Shakhty Pedagogical Institute has been devoted to this problem, and the result is the apparatus described below for studying light scattered by the atmosphere. This apparatus is an improved and completely automated version of the original device described in [2].

This problem has been raised earlier by many researchers, as is apparent from the instruments created by Fesenkov, Dietze, Myukhkyurya, Stepanov, et al. [3, 4, 5]. It is difficult to evaluate the advantages and disadvantages of the existing instruments, since all of them, just as the one described here is, are chiefly designed for a specific area of research. Our main aim in designing the instrument was to have as complete as possible automation of the measurement process itself and to obtain the most complete information possible about scattered light. Our apparatus measures the total intensity of scattered light and the intensity of the two mutually perpendicular components of partially scattered light in relative units, the degree of polarization, and the position of the plane of polarization, i.e., three of the four Stokes parameters. The fourth Stokes parameter — the ellipticity of polarization — is either very small or entirely absent for light scattered by the atmosphere [6], so we did not provide for its measurement. Since the measurement process is automated, measurements are made at specific points in the sky — at points on the almucantars with zenith distances of 90, 85, 80, 75, 70, 65, 60, 50, 40, 30, 20, and 10° at an angular distance of 20° from one another. The angular distances of the points on the azimuth are read from the solar vertical circle. The measurement time for one almucantar (17 points) is 3 min. Thus, the entire sky is covered in one series of measurements (all almucantars). In addition, measurements can be made at 25 points on an arbitrarily selected vertical circle that are at an angular distance of 5 or 10° from one another, according to the above zenith distances of the almucantars. Measurements in one vertical circle take 4 min. Measurements at any point in the sky at 10-sec intervals are also possible.

The apparatus consists of a photometer, a photocurrent amplifier, a recorder, an automatic device, and a servo system. The light from the observed region of the sky goes through a system of diaphragms which form a 30' cone to a uniformly rotating analyzer (a Franck-Ritter polarizing prism, unary, truncated), and then, after passing through a depolarizer (milk glass), it enters an FEU-35 photomultiplier. The photocurrent is amplified by a dc amplifier

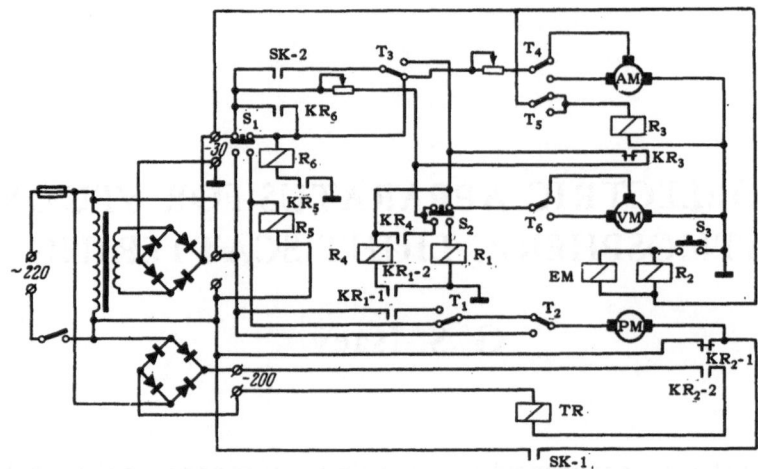

Fig. 1. Schematic diagram of automatic device.

and recorded by an EPP-09 electronic potentiometer. The azimuthal and vertical scales were taken from a TG-1 theodolite.

The automatic device (Fig. 1) is designed to ensure operation of the instrument on the almucantars, on the vertical circle, at a point. When tumbler switches T_1, T_2, and T_3 are in the positions shown in Fig. 1, the apparatus is set for almucantar measurements. A dc voltage of 30 V flows through the normally closed contacts of terminal switch S_1 and reversing tumbler switch T_4 to the azimuth motor AM, which begins to turn the photometer about the vertical axis. When the photometer has turned 20°, switch S_1 opens the circuit of the motor AM and simultaneously actuates the analyzer motor PM through T_1 and T_2 and the normally closed contacts KR_2-1. Motor PM begins to turn the analyzer, and when it has made a complete revolution, switch S_3 closes the circuit of relay R_2 and the electromagnet EM. Relay R_2 breaks the circuit of motor PM and simultaneously actuates the time relay TR. The electromagnet EM closes the shutter of the photometer. One second after its actuation, the time relay closes sliding contacts SK-2 briefly, which actuates motor AM and locking relay R_6. This closes contacts KR_6, shunting the now open contacts S_1, and this lasts until S_1 returns to its original position, having opened the circuit of relay R_5 and thereby having opened the circuit of relay R_6. Motor AM begins to turn, switching S_1 to its original position, and, after having turned the photometer 20° about the vertical axis, it again switches S_1. By then, the time relay TR has closed contacts SK-1, actuating motor PM. The analyzer begins to turn and opens S_3, opening the circuits of the electromagnet EM and relay R_2, which opens the circuit of the time relay and actuates the motor through KR_2-1. The electromagnet opens the shutter of the photometer. After the analyzer has turned 360°, the shutter is closed and the instrument moves to the next point. When the photometer makes a complete revolution about the vertical axis, the reversing tumbler switch T_4, and T_5, which is connected to it, are switched. Switch T_4 reverses motor AM, and at the moment of switching, T_5 opens the circuit of relay R_3 for an instant. When relay R_3 is actuated, the normally closed contacts KR_3 shunt the contacts of S_2, and the 30 V goes to the vertical motor VM and locking relay R_4. Contacts KR_4 shunt the upper contacts of S_2, and motor VM begins to turn the photometer about the horizontal axis. When the photometer has turned 5 or 10°, depending upon the original position, S_2 is switched to the second position, opening the circuits of motor VM and relay R_4. Thus, the photometer is moved to the next almucantar, where it again begins measurements at points 20° from one another.

For measurements on a vertical circle, T_1 and T_3 are moved to their upper positions, switching the sliding contacts SK-2 of the time relay in parallel with the normally closed

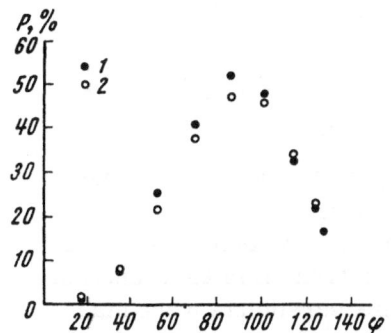

Fig. 2. Degree of polarization versus scattering angle: 1) for points to the west of the solar vertical circle; 2) to the east.

contacts of S_2 and slightly changing the circuit of the analyzer motor PM. Then the 30 V goes through the normally closed contacts of S_2 and the reversing switch T_6 to motor VM. When the photometer has turned 5 or 10° about the horizontal axis, S_2 is switched to the second position, opening the circuit of motor VM and actuating relay R_1, whose contacts KR_1-1 close the circuit of the analyzer motor PM. After the analyzer has turned 360°, S_3 actuates the electromagnet and relay R_2. The electromagnet closes the shutter. Relay R_2 actuates the time relay, whose sliding contacts SK-2 after 1 sec actuate locking relay R_4 and motor VM. Relay R_4 keeps its contacts KR_4 closed until S_2 returns to its original position. The photometer again turns 5 or 10° about the horizontal axis. By this time, contacts SK-1 of the time relay have closed, actuating the analyzer motor PM. Beginning to turn, the analyzer switches S_3, opening the circuits of relay R_2 and the electromagnet. The shutter is opened, and the time relay is cut off. After the analyzer has turned 360°, the entire process is repeated.

For measurements at a point, switch T_2 is placed in its lower position and a special tumbler switch (not shown) cuts off motors AM and VM. Then 220 V goes through T_2 and contacts KR_1-1 to the analyzer motor PM. After the analyzer has made a complete revolution, S_3 is closed, actuating the electromagnet and relay R_2. The electromagnet closes the shutter, and relay R_2 disconnects the motor PM and actuates the time relay. After 6 sec, contacts SK-1 of the time relay close, actuating the analyzer motor. As the analyzer begins to turn, S_3 is opened, disconnecting the electromagnet and relay R_2. The electromagnet opens the shutter, and relay R_2 cuts off the time relay. Then the process is repeated.

After the measurements, the recording tape is processed, and the maximum and minimum components of partially polarized light are read from it with a millimeter rule. Then the total intensity ($I_{max} + I_{min}$), the amount of polarized light ($I_{max} - I_{min}$), and the degree of polarization are determined. The scattering angles are calculated and graphs are plotted from the azimuth and height of the observed point and the height of the sun, which is taken from Sharonov's tables. Since all of the measured values are in relative units, the main sources of error are nonlinearity of the amplifier and the direct measurements when reading the recording tape with a ruler.

The amplifier was checked beforehand and was found to be practically linear within the selected voltage range, which was completely determined by the width of the recording tape. The absolute error in measurement with the ruler does not exceed one-half of the smallest division, i.e., 0.5 mm. The absolute error is 1 mm for the sum and the difference of the components. The relative error is greatly dependent upon the numerical values themselves, so it varies within wide limits, especially for the difference. Because of·this, the error in determining the degree of polarization is not constant from measurement to measurement and is greatly dependent upon the intensity difference. The absolute error in determining the degree of polarization varies from fractions of a percent to 4-5%, which is entirely permissible for the conventional scales of the graphs. In general, the error decreases as each intensity or their difference increases. In doubtful cases, the absolute and relative errors can be calculated for each specific value. The position of the plane of polarization was determined by the angular distance from the beginning of rotation of the analyzer to the maximum or minimum of the curve [5]. The angular distances of the observation direction are set automatically with accuracy to 1°, which is entirely sufficient when tables are used to determine the height and azimuth of the sun. The azimuthal and vertical scales of the photometer can be read with accuracy to 1'.

The servo system, whose circuit was proposed by I. Ya. Badinov [7], is used to put the axis of the instrument into the plane of the solar vertical circle at zero azimuth.

The apparatus was built in the workshops of the Shakhty Pedagogical Institute. At the present time, measurements are being made according to a specific program. A certain amount of experimental data have been accumulated. The results will be presented later. An example of the results is shown in Fig. 2, which gives the degree of polarization as a function of the scattering angle for the solar almucantar with z = 65°. The measurements were made at 1138 hours local time on 9 November 1966. As can be seen, the polarization is not completely symmetric relative to the solar vertical circle. This asymmetry, which was observed for other almucantars, can be explained by the fact that the optical properties of the atmosphere were not the same in all directions.

LITERATURE CITED

1. D. G. Stamov, "On the possibility of polarimetric determination of turbidity in different directions," in: Actinometry and Atmospheric Optics [in Russian], Leningrad (1961).
2. G. S. Isaev, "A semiautomatic polarimeter for almucantar measurements of the polarization of light scattered in the atmosphere" (in press).
3. V. G. Fesenkov, "The two-channel polarimeter and its use in atmospheric optics and astrophysics," Astron. zh., Vol. 36, No. 6 (1959).
4. G. Dietze, Z. Meteorolog., Vol. 14, Nos. 7-9 (1960).
5. V. I. Myukhkyurya, "A general-purpose photoelectric photometer," Trudy GGO, No. 93 (1959).
6. V. G. Fesenkov, "On the presence of elliptical polarization in the daylight sky," Astron. zh., Vol. 37, No. 5 (1960).
7. I. Ya. Badinov, "On automatic sun tracking in field studies of the optical properties of the atmosphere, "Nauchn. soobshch. Inst. geol. i geogr. Akad. Nauk Lit. SSR, No. 13 (1962).

RATIONALITY OF CHARACTERISTIC FOR OPTICAL STATE OF THE ATMOSPHERE

I. N. Yaroslavtsev and V. V. Sakhanov

A great many of the proposed characteristics for determining the optical state of the atmosphere have little suitability for recognizing in time the approach of a synoptic air mass [1–3]. Kh. Myurk, in attempting to find the best, most practically suitable characteristic, proposed the use of a rationality factor r. The main criteria of rationality are high sensitivity to atmospheric transmittance and little dependence upon the height of the sun. Then r is determined by the formula

$$r = \frac{1}{\Pi}\frac{\partial \Pi}{\partial S} : \frac{1}{\Pi}\frac{\partial \Pi}{\partial m},$$

where Π is the transmittance characteristic, and the derivatives $\partial \Pi / \partial S$ and $\partial \Pi / \partial m$ determine its dependence upon the density of solar radiation flux and atmospheric mass (height of the sun).

According to Myurk's data, his parameter B and Makhotkin's turbidity index N are the best from this point of view.

According to the International Meteorological Association [5, 6], however, the Bouguer transmission coefficient p and the Linke turbidity factor T have gained the widest use in the USSR, although the rationality factors of these characteristics are, according to Myurk [7], considerably lower than the rationality factors of B and N. Comparison under identical optical conditions of the transmittance characteristics in Table 1 is therefore of decided interest.

The characteristics were calculated from observations in Ryazan' of direct solar radiation by actinometers designed by Yu. D. Yanishevshii [8]. An adapter with an IKS–3 plane glass filter, whose transmission coefficients were determined on an IKS–14 spectrometer, was installed on the actinometer to separate the infrared radiation (0.8–3.0 μ) from the integrated flux (0.3–3.0 μ). The reduction factors of the filter were calculated from the modern atmospheric model proposed by Avaste, Moldau, and Shifrin [9]. The conversion factors of the actinometer with and without the filter were determined by comparison with Ångström pyrheliometers No. 250 (USSR standard) and No. 569 (Ryazan' standard), the constant of which was obtained by comparison in Sweden with the European standard, pyrheliometer No. 158. The values of direct solar radiation, which were obtained in the international pyrheliometric scale of 1956, were reduced to the mean distance between the earth and the sun.

The IKS–3 filter made it possible to isolate the following spectral regions of direct solar radiation: integrated, from 0.3 to 3.0 μ; visible and ultraviolet, from 0.3 to 0.8 μ; and infra-

TABLE 1. Characteristics of Atmospheric Transmittance and Basic
Formulas for Calculating Them

Characteristic	Basic formula	Source
Bouguer transmission coefficient, p ...	$P_M = \dfrac{\log S - \log S_0}{M}$	[6, 10]
Linke turbidity factor, T	$T_M = \dfrac{\log P_M}{\log q_M}$	[6, 10]
Myurk parameter, B............	$S = S_0 \cdot p_1^M \cdot M^{B \cdot M}$	[3, 11]
Makhotkin turbidity index, N.......	$N = \dfrac{M^*}{M}$	[4]

N o t e . S_0 is the solar constant; M the number of masses of the real atmosphere for
which direct radiation S was measured; q the transmission coefficient of an ideal atmosphere; p_1 the transmission coefficient of the real atmosphere at M = 1; M^* the
number of masses reduced to a "normal" atmosphere assuming $S = S^*$; S^* the flux density of direct radiation in a normal atmosphere. Makhotkin takes as a normal atmosphere the model given by the formula $S^* + 0.5(S - 0.8)^3 = 1.41 - 1.11 \log M^*$.

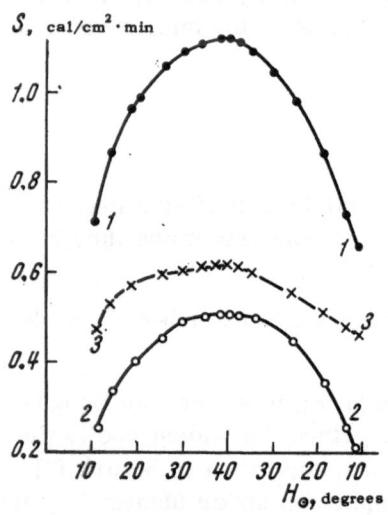

Fig. 1. Daily variation of flux density of solar radiation on 2 April
1966: 1) in wavelength intervals
from 0.3 to 3.0 μ ; 2) from 0.3 to
0.8 μ ; 3) from 0.8 to 3.0 μ . $H_\odot$
is the sun's height and S is the
intensity of direct solar radiation.

red, from 0.8 to 3.0 μ . These data were used to calculate the above characteristics of atmospheric transmittance. To eliminate the effect of the sun's height,
p and T were reduced to a sun height of 30° [6]. Some
of the calculation results are given in the tables and
figures. For example, Fig. 1 shows the daily variation
of the flux density (intensity) of solar radiation in a
cloudless sky that was most typical for the middle
geographical lattitudes of the European part of the USSR
when turbidity increased during the afternoon hours.

The variability of the integral characteristics of
atmospheric transmittance during this period can be
seen in Fig. 2. The daily variation of the characteristics
corresponds to the degree of turbidity.

Five days with different degrees of turbidity (clear,
hazy, foggy, cloudy) were arbitrarily selected to show
the sensitivity of the characteristics to the transmittance of the real atmosphere. The arithmetic means
of the characteristics (Π), their variability ($\Delta\Pi$), degree of variability ($\Delta\Pi / \Pi$), and weighted means for the
day are shown in Table 2.

Similar calculations were made for a number of
real [4, 9, 10] atmospheric models (Table 3) for different
degrees of turbidity.

The data indicate the following.

1) In order of decreasing sensitivity to transmittance, the characteristics are: the turbidity index N, the turbidity coefficient A (about which more will be said later), the parameter
B, the turbidity factor T, and the transmission coefficient p.

2) The sequence of characteristics in order of increasing sensitivity to sun height is:
p, B, A, T, and N.

TABLE 2. Sensitivities of Characteristics as Functions of Solar Radiation (Atmospheric Transmittance)

Date	Number of observation series	p	Δp	$\frac{\Delta p}{p}$, %	T	ΔT	$\frac{\Delta T}{T}$, %	B	ΔB	$\frac{\Delta B}{B}$, %	N	ΔN	$\frac{\Delta N}{N}$, %	A	ΔA	$\frac{\Delta A}{A}$, %
17 November 1965	5	0.810	0.003	0.4	2.08	0.04	1.7	0.049	0.001	2.0	0.57	0.01	1.8	0.77	0.02	2.6
18 November	5	0.809	0.003	0.4	2.12	0.04	2.1	0.050	0.001	2.0	0.59	0.02	2.5	0.80	0.02	2.5
20 October	15	0.713	0.032	4.5	3.40	0.48	14.0	0.081	0.012	15.4	1.11	0.20	18.4	1.55	0.32	20.6
24 March 1966	13	0.692	0.051	7.4	3.70	0.70	19.0	0.092	0.020	22.0	1.33	0.47	35.0	1.80	0.55	31.0
2 April	15	0.735	0.016	2.2	3.07	0.20	6.5	0.074	0.005	6.8	0.99	0.09	9.1	1.34	0.14	10.4
Weighted mean		0.732	0.027	3.7	3.14	0.37	11.8	0.076	0.010	13.0	1.03	0.20	19.4	1.41	0.27	19.2

TABLE 3. Sensitivities of Characteristics as Functions of Sun Height

M	Makhotkin model					$\tau_0 = 0.2$, $\omega = 0.5$ cm					$\tau_0 = 0.3$, $\omega = 2.1$ cm					$\tau_0 = 0.5$, $\omega = 0.5$ cm				
	p	T	B	N	A	p	T	B	N	A	p	T	B	N	A	p	T	B	N	A
1	0.728	3.17	0.075	1.0	1.50	0.795	2.30	0.054	0.646	1.0	0.704	3.51	0.085	1.16	1.70	0.65	4.35	0.106	1.56	2.2
1.5	0.732	3.12	0.075	1.0	1.44	0.789	2.38	0.055	0.674	1.0	0.704	3.51	0.084	1.20	1.73	0.63	4.59	0.108	1.72	2.3
2	0.734	3.10	0.073	1.0	1.43	0.785	2.42	0.055	0.686	1.0	0.700	3.57	0.084	1.24	1.70	0.63	4.59	0.112	1.82	2.4
2.5	0.735	3.08	0.073	1.0	1.38	0.782	2.46	0.056	0.704	1.0	0.697	3.62	0.085	1.27	1.74	0.62	4.79	0.114	1.90	2.5
3	0.732	3.12	0.071	1.0	1.36	0.782	2.46	0.056	0.724	1.0	0.693	3.67	0.085	1.30	1.73	0.61	4.99	0.115	2.00	2.7
4	0.732	3.12	0.071	1.0	1.34	0.779	2.50	0.056	0.740	1.0	0.682	3.83	0.086	1.35	1.80	0.60	5.05	0.117	2.14	—
5	0.732	3.12	0.069	1.0	1.33	0.776	2.54	0.056	0.755	1.0	0.678	3.89	0.087	1.40	—	0.59	5.19	0.120	2.28	—
6	0.732	3.12	0.068	1.0	1.35	0.769	2.62	0.057	0.760	1.0	0.670	4.00	0.088	1.45	—	0.59	5.19	0.119	2.35	2.4
Π	0.732	3.12	0.072	1.0	1.40	0.782	2.45	0.056	0.710	1.0	0.692	3.72	0.086	1.30	1.73	0.61	4.84	0.114	1.97	
$\frac{\Delta\Pi}{\Pi}$	0.1%	0.4%	3%		3.6%	0.7	3%	1.1%	5%		1.7%	4.3%	1.2%	6.2%	1.4%	3%	5%	3%	11%	4%

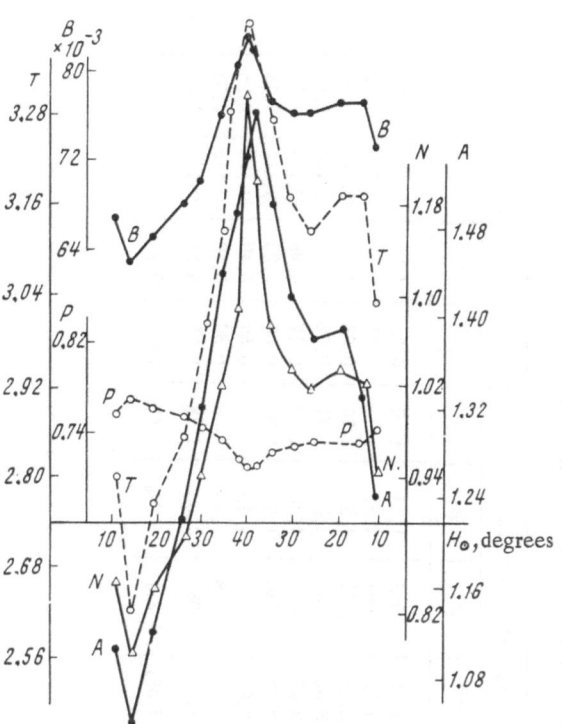

Fig. 2. Daily variation of characteristics of integral transmittance.

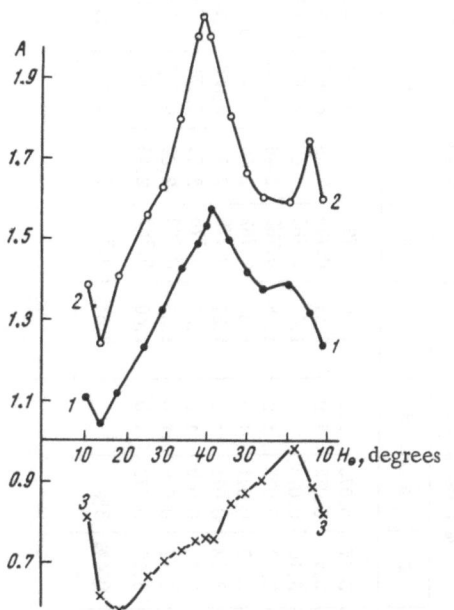

Fig. 3. Variation of turbidity coefficient A on 2 April 1966: 1) 0.3–3.0 μ; 2) 0.3–0.8 μ; 3) 0.8–3.0 μ. $H_\odot$ is the sun height.

3) According to our calculations, the rationality factors for these characteristics have the following most probable values: A = 8.7–11.3, B = 6.2–10.5, T = 3.7–6.7, N = 3.5–4.3, and p = 2.7–5.2.

The spectral transmission coefficients of the atmosphere must be known in order to solve a number of problems in astrophysics, geophysics, meteorology, and instrument building. Of all of the characteristics examined, only the Bouguer transmission coefficient p and the Linke turbidity factor T can be used directly for these purposes. But they also have important shortcomings, as can be seen from the following example. From the flux densities of direct solar radiation measured on 17 November 1955 we calculated the mean values of p and T for the integrated, infrared, and visible radiations:

	p	T
Integrated	0.848	2.04
Infrared	0.920	5.18
Visible	0.800	1.72

On this day, the atmosphere was more transparent to infrared, but the turbidity factor had lower values only for the visible region. The results were not comparable. We therefore used the turbidity coefficient A. It was calculated as was the turbidity index N, i.e., A = M*/M, where M is the number of atmospheric masses at the moment at which intensity S was measured and M* is the number of masses for a slightly turbid model [9] for which the optical thickness $\tau_0 = 0.2$, the amount of water $\omega = 0.5$ cm, and S = S* (S* is the radiation intensity in this model). The advantages of this characteristic are apparent from the tables and Figs. 2 and 3.

Summary

1. The turbidity coefficient A allows atmospheric transmittance to be assessed quickly, fairly accurately, and without tedious preparation.

2. The coefficient A characterizes atmospheric turbidity. It indicates the factor that must be used to change the number of masses of the standard atmosphere to make its optical properties correspond to reality.

3. The coefficient A allows transmittance to be compared in various spectral regions.

LITERATURE CITED

1. K. Ya. Kondrat'ev, Actinometry [in Russian], Leningrad (1965).
2. S. I. Sivkov, "Comparison and critique of various methods of reducing atmospheric transmission coefficients to a unit mass," Trudy GGO, No. 14(76), p. 35 (1949).
3. Kh. Myurk, "On a new formula for radiation intensity and new characteristics of atmospheric transmittance. Research in atmospheric physics." Trudy Inst. fiziki i astronomii (IFA), Akad. Nauk Est. SSR, Tartu, No. 1, p. 7 (1959).
4. L. G. Makhotkin, Izv. Akad. Nauk SSSR, seriya geofiz., No. 5 (1957).
5. S. I. Sivkov, "Generalization of empirical relations between solar-radiation intensity, sun height, and atmospheric transmittance," Trudy GGO, No. 115, p. 95 (1960).
6. Instructions for Determining Atmospheric-Transmittance Characteristics for Actinometry Departments of Hydrometeorological Observatories [in Russian], Leningrad (1965).
7. Kh. Myurk, "On the rationality of Makhotkin's turbidity index. Research in atmospheric physics." Trudy IFA Akad. Nauk Est. SSR, Tartu, No. 1, p. 26 (1959).
8. Yu. D. Yanishevskii, Actinometric Instruments and Observation Methods [in Russian], Moscow (1957), p. 203.
9. O. Avaste, Kh. Moldau, and K. S. Shifrin, "Spectral distribution of direct and scattered radiation. Research in atmospheric physics." Trudy IFA Akad. Nauk Est. SSR, Tartu, No. 3, p. 23 (1962).
10. S. I. Sivkov, "Solar radiation attenuation in an ideal atmosphere," Trudy GGO, No. 169, p. 66 (1965).
11. Kh. Myurk, "Nomograph for calculating and reducing certain characteristics of atmospheric transmittance. Research in atmospheric physics." Trudy IFA Akad. Nauk Est. SSR, Tartu, No. 1, p. 15 (1959).

METHOD FOR STUDYING TWILIGHT PHENOMENA

V. G. Fesenkov

1. Many papers have been devoted to the investigation of twilight phenomena, but they have been mainly of a descriptive nature. Conclusions about the optical properties of the upper atmosphere, which, it would seem, should be the principal goal of twilight research, are usually drawn in a highly elementary manner, on the basis of rather arbitrary simplifying assumptions. The optical properties of twilight are incomparably more complex than those of the daytime sky.

In the case of the daytime sky, in fact, in most cases we can consider the earth a plane, ignore atmospheric refraction, the finite dimensions of the solar disk, and the height distribution of atmospheric ozone, and find the transmission coefficient and scattering indicatrix from direct observations on the solar almucantar. But in considering twilight phenomena, we must make special allowance for all of these factors, and also a number of others, such as the variation of the scattering indicatrix with height, so-called refractive dispersion, i.e., the divergence of solar rays that have been unequally refracted at different heights from the earth's surface, and, in particular, higher-order scattering, which is produced mainly by the bright parts of the high-level twilight segment, which illuminates the other, lower atmospheric layers. This effect is especially great near the zenith, which makes twilight observations in this region rather fruitless.

With modern computer technology, however, it quite possible and necessary to develop a suitable method for studying twilight phenomena, with primary attention given to optical sounding of the atmosphere up to 150-160 km, and also of the dust cloud.

2. It should be noted, first of all, that even before sunset, the atmosphere begins a gradual transition to the twilight state. For example, as our calculations have shown, the brightness of the sky at the zenith when the sun is in a high position is determined mainly by the lowest and densest air layers. But as the sun approaches the horizon, the contribution of the surface layers decreases rapidly, so that at sunset, when the atmospheric mass is about 40 units, the maximum zenith illumination is caused by layers at a height of about 20 km. On the side of the sky opposite the sun, the shadow contour appears and rises rapidly, and this is clearly dependent upon the height distribution of atmospheric ozone. This distribution can be determined by comparing observations made above the shadow boundary inside and outside the ozone absorption band.

According to our observations, which were made in 1948 under exceptional atmospheric conditions in the Sary Ishik Ottrau Desert, to the south of Lake Balkhash, the shadow height depends upon the sun's depression as follows:

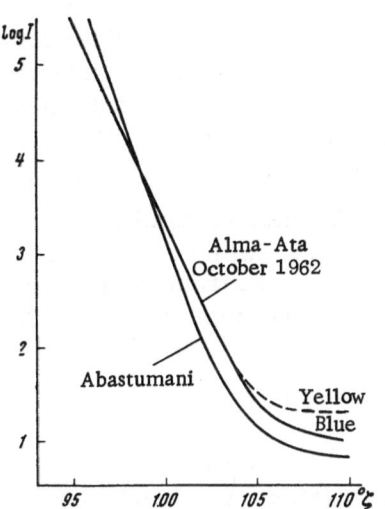

Fig. 1. Averaged twilight curves for Alma-Ata and Abastumani.

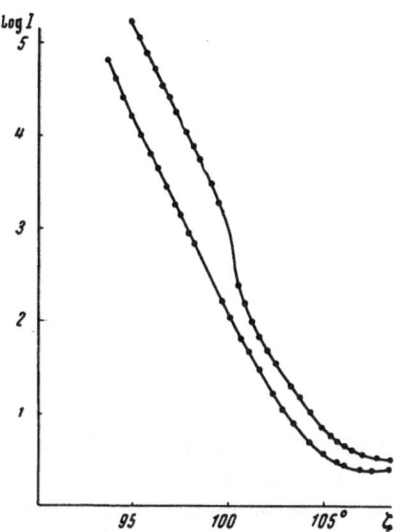

Fig. 2. Twilight curves for symmetric points on solar vertical circle at zenith distances of 70° (blue filter, λ = 424 mμ), Alma-Ata, 4 October 1962.

Depression of sun	1°	2°	3°	4°	5°
Shadow height	2°2	4°1	6°9	10°5	17°2

When the sun's depression is 6° or more, the shadow height already considerably exceeds 20°. Hence, it follows that the brightness of the twilight sky on the antisolar vertical circle at a zenith distance of 70° is entirely determined by higher-order scattering and not by direct radiation.

3. Now let us consider the common twilight phenomena — the total effect of primary scattering in fairly high layers, higher-order scattering, i.e., the illumination in the lower layers caused by the bright twilight segment, and, finally, the nighttime sky background, which is also complex.

This third component begins to be appreciable when the sun is 14° or so below the horizon, and it manifests itself most strongly in the yellow region of the spectrum. Moreover, it is interesting to note that the gradient of the twilight curve (log I, ζ) appears to vary from place to place. But this requires careful checking. Figure 1 shows averaged twilight curves obtained at Alma-Ata (Kamenskoe Plateau) and Abastumani Observatory for a point on the solar vertical circle at a zenith distance of 70°. As can be seen, the twilight brightness decreases somewhat more slowly at the first location. This indicates the extreme desirability of simultaneous observations at different locations with instruments of the same type that have been carefully compared with one another.

As we showed earlier [2], in order to eliminate higher-order scattering, which is difficult to evaluate theoretically, it is advisable to make parallel observations at two symmetric points on the solar vertical circle at about 20° above the horizon. At the second symmetric point, which is on the side opposite the sun, the observed brightness is due exclusively to higher-order scattering when the sun is more than 6° below the horizon. At this point, the twilight curve is distinguished by rather high regularity (Fig. 2). At the first point, where the brightness is due mainly to primary scattering, i.e., to directly illuminated layers at great height, characteristic irregularities are often observed. These irregularities usually disappear after the sun is more than 10° below the horizon.

These irregularities are for the most part caused by distinct heterogeneities of the upper atmosphere — interlayers that appear above 90 km, which contain material of meteoric origin. It would, therefore, be especially interesting to make systematic observations of twilight phenomena when meteor showers and noctilucent clouds are present.

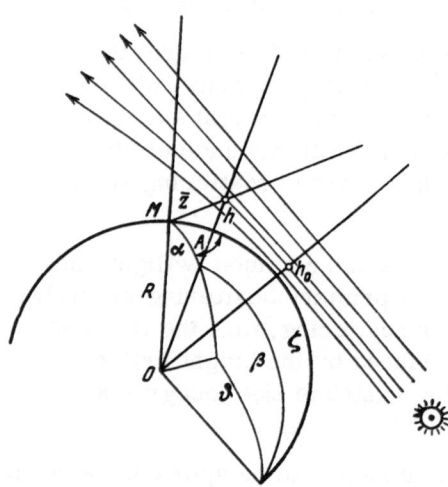

Fig. 3. Distribution of spherical coordinates.

Even higher layers evidently have considerably greater regularity with respect to the distribution of scattering material. According to preliminary data [2], this distribution is characterized by a scale height of 18 km, which is much greater than the scale height of an atmosphere of ordinary gaseous composition.

4. Let us consider the mathematical expression for twilight brightness at a point with coordinates (z_0, A_0), where the azimuth is read from the solar vertical circle. As was noted above, this brightness consists of three components:

$$I = I_1 + I_2 + I_3,$$

where

$$I_1 = L\omega \frac{f^0(\vartheta)}{\sin\vartheta} p^{\sec z_0} \int_{h_0\,\min}^{\infty} \mu(h)\,dh_0 \int_{\beta-\rho_0}^{\beta+\rho_0} je^{-\tau_{oz}}e^{-\tau_a}f\,d\beta$$

and

$$I_2 = \int_0^{\frac{\pi}{2}+\Delta} dz \int_0^{\pi} B(z, A) f(\vartheta) \varphi(z, z_0) \sin z\, dA,$$

where

$$f = \frac{d(n_0 r_0)}{dh_0} \cdot \frac{r}{n_0 r_0} \cdot \frac{dh_0}{dh};$$

$$\frac{\tau}{\tau_\infty} = 2\frac{\displaystyle\int_0^{\varphi_{\max}} F(h)\, r\, \mathrm{cosec}\,\eta\, d\varphi}{\displaystyle\int_0^{\infty} F(h)\, dh};$$

$$\varphi(z, z_0) = \frac{\mu}{k} \cdot \frac{p^{m_z} - p^{m_{z_0}}}{m_{z_0} - m_z} \cdot m_{z_0}.$$

Here, L is the solar-radiation flux per steradian; ω the spatial angle of the sun; $f^0(\vartheta)$ the scattering indicatrix reduced to high atmospheric layers; $f(\vartheta)$ the overall scattering indicatrix, which is found from direct observations; $p^{\sec z_0}$ the transmission coefficient of the entire atmosphere in the direction to a given point; $\mu(h)$ the scattering function, which is to be determined by analyzing twilight phenomena and is characteristic for higher layers; τ_{oz} and τ_a the ozone and aerosol optical thicknesses, which are related to the ozone and aerosol height distributions F(h); f the refractive-dispersion factor, which is determined by the change in $n_0 r_0$ (n_0 is the refractive index at height h_0, the closest the light trajectory comes to the earth's surface); B(z, A) the brightness of the primary twilight segment, which for the point (z, A) is represented by an expression similar to that for I_1; and $\varphi(z, z_0)$ is the atmospheric-mass function, which depends only upon zenith distance. The remaining symbols are explained in Fig. 3.

It is quite natural to assume that the gaseous structure of the atmosphere, which causes refractive phenomena, is fairly well known, since it is mainly connected with the lower layers. The refractive-dispersion factor can therefore be calculated, as has been done on electronic computers with the assistance of the Astronomical Council of the Academy of Sciences of the USSR. The obtained tables represent the light trajectories at various perigees h_0 above the

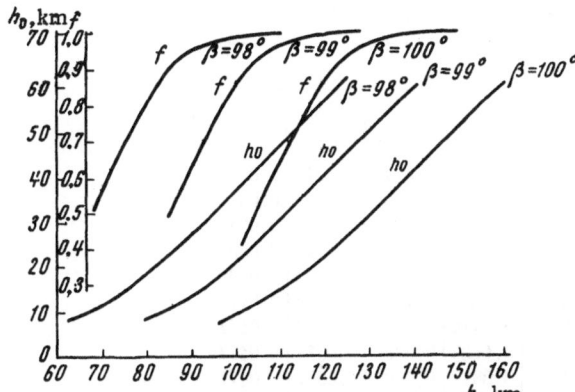

Fig. 4. Relationship between heights h_0 and h, and the refractive-dispersion factor f as a function of h_0 and β.

earth's surface. They are expressed in polar coordinates (r, φ) relative to the center of the earth and give as functions of h_0 the heights h of the scattering elements of the atmosphere and the refractive-dispersion factors f for each β value, which is the angle between the sun and the scattering body.

A small part of these tables is illustrated in Fig. 4. As can be seen, merely one refractive nonuniformity and the spread of light trajectories caused by it can reduce the brightness of the illuminated element of the atmosphere by a factor of up to 2.5 or even 3.

On the other hand, the height distribution of ozone and aerosols, which determines the absorption and extinction of solar radiation, is associated with comparatively low layers, which reach up to approximately 40-50 km. This distribution must be determined independently, for example, by observations of artificial satellites when they enter the earth's shadow. In fact, when an artificial satellite enters the earth's shadow, it is illuminated by rays from an extremely oblate solar disk, which shines through the lower atmospheric layers. And these layers, which are 40-50 km thick, seem, from the height of the satellite (about 1500 km) and at the corresponding distance of the horizon (tens of thousands of kilometers), to have the same angular dimensions as the solar disk. Such observations have been made by V. S. Matyagin at the Astrophysical Institute, Alma-Ata. The results are presented in [3]. At the present time, for systematic observations of this kind, a special device is being built at the Astrophysical Institute for accurate recording in different spectral regions of the decrease in the brightness of a satellite for the 10-12 sec that it is visible. Observations inside and outside the ozone absorption band permit reliable determination of the height distribution of atmospheric ozone and aerosols.

In this way, we can determine all of the factors under the integral sign in the expression for I_1 with the exception of $\mu(h)$, which characterizes the optical properties of the upper atmosphere and must be determined from twilight observations. For this, however, we must deduct from the observed total twilight brightness not only component I_3 but also I_2, i.e., the higher-order scattering, which it is more correct to call the tropospheric component. Without such reduction, the interpretation of twilight phenomena has no real meaning.

5. Let us show, first of all, that twilight observations near the zenith are rather useless, because they are greatly affected by higher-order scattering which is impossible to eliminate. For this, it is sufficient to calculate I_1 and I_2 for the most probable atmospheric structure at great heights, using the approximate expression for primary twilight (ignoring refraction and the finite dimensions of the solar disk):

$$B(z, A) \sim \frac{f(\vartheta)}{\sin \vartheta} \int_{h_{0\min}}^{\infty} e^{-\frac{h}{H}} [1 - e^{-A(h_0 - B)^2}] \, dh_0$$

at $H = 20$ km, $A = 0.004$, and $B = 9$ km.

We calculated the isophots of the primary twilight segment for the entire sky in arbitrary units using this formula. The corresponding brightnesses are shown in the table for a solar zenith distance of 100°, when the twilight segment is sufficiently far from the tropo-

Brightnesses of Primary Twilight Segment at $\zeta = 100°$

z	\multicolumn{10}{c}{A}									
	0°	10°	20°	30°	40°	50°	60°	70°	80°	90°
90°	211.7	123.6	51.41	23.50	10.84	4.98	2.93	0.867	0.301	0.0593
88	112.9	72.5	32.95	15.0	6.98	3.33	1.53	0.615	0.229	0.0559
85	46.8	33.7	16.72	8.3	4.02	2.01	0.917	0.407	0.171	0.0536
80	14.3	11.46	6.56	3.40	1.80	1.00	0.487	0.244	0.119	0.0504
75	5.52	4.65	3.06	1.80	1.03	0.580	0.313	0.172	0.0962	0.0492
70	2.65	2.36	1.69	1.03	0.650	0.370	0.228	0.1384	0.0838	0.0478
60	0.846	0.780	0.629	0.455	0.316	0.215	0.1474	0.102	0.0709	0.0477
50	0.373	0.361	0.338	0.259	0.197	0.146	0.1125	0.0830	0.0634	0.0477
40	0.197	0.189	0.182	0.155	0.125	0.1083	0.0817	0.070	0.0581	0.0477
30	0.1186	0.1136	0.1093	0.0985	0.0891	0.0786	0.0712	0.0618	0.0544	0.0476
20	0.0853	0.0836	0.0807	0.0760	0.0716	0.0672	0.0621	0.0562	0.0521	0.0476
10	0.0641	0.0622	0.0604	0.0590	0.0578	0.0560	0.0530	0.0512	0.0491	0.0475
0	0.0475	0.0475	0.0475	0.0475	0.0475	0.0475	0.0475	0.0475	0.0475	0.0475

sphere above the observer's horizon. If, under these conditions, we consider each element of the primary twilight segment as an independent and distant light source, we can calculate I_2 at the zenith, and also for the two 70° symmetric points on the solar vertical circle. Then we can compare the obtained brightnesses directly with I_1. Allowance for polarization of the twilight segment does not appreciably change the results (see below):

$$\zeta = 100°$$

z	70°(toward sun)	0°	70° (away from sun)
I_1	2.65	0.0475	0.000
I_2	0.118	0.0403	0.0511
I_2/I_1	4.4%	84.8%	∞

We shall show, in addition, that the illumination that occurs at the zenith is caused by different parts of the primary twilight segment, which correspond to entirely different heights. For example, the relative contributions Φ of various almucantars and the corresponding effective heights h_{eff} are:

z	Φ	h_{eff}, km	z	Φ	h_{eff}, km	z	Φ	h_{eff}, km
90°	77.91	36.5	75°	18.14	81.2	40°	1.78	105.4
88	81.40	49.7	70	10.92	87.1	30	1.20	109.05
85	66.34	60.7	60	4.88	95.3	20	0.87	112.2
80	33.93	72.7	50	2.83	101.1	0	0.00	118.6

As can be seen from these data, the brightness of the twilight sky at the zenith, even when the sun is in an intermediate position, is a rather complex mixture of illuminations from atmospheric layers at a variety of heights plus a certain amount of primary scattering. Conversely, under these same conditions, the contribution of the brighter parts of the primary twilight segment is comparatively small, and, when simultaneous observations are made at the second symmetric point, it can be completely eliminated.

In fact, if at the first symmetric point,

$$I = I_1 + I_2 + I_3$$

and at the second

$$I' = I'_2 + I'_3 \qquad (I'_1 = 0),$$

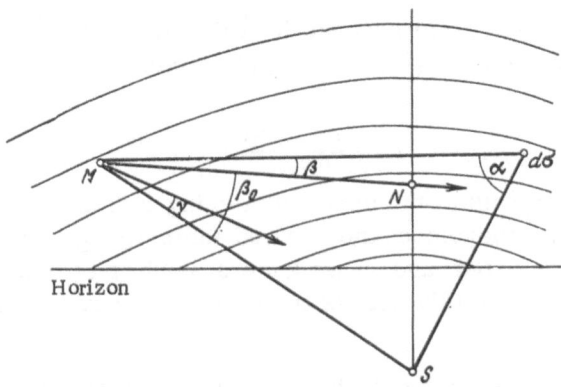

Fig. 5. Primary twilight segment and illumination caused by it at a point not on the solar vertical circle.

and we determine theoretically the ratio $K = I_2/I_2'$, which is chiefly dependent upon the scattering indicatrix, we find

$$I_1 = I - I_3 - KI_2' .$$

Note that the conversion factor K is easily determined from the system of isophots of the primary twilight segment. When $\zeta = 100°$, for example, $K = 2.388$, if we ignore the natural polarization of the primary segment. When polarization is taken into account, we obtain $K = 2.301$.

In deep twilight, shortly before nightfall, when scattering by higher atmospheric layers appears, the brightness observed at $z = 70°$, even on the side of the sun, is in fact due to higher-order scattering, since I_1 becomes negligible in comparison with I_2. Thus, the ratio method is no longer applicable, but observations at the two symmetric points on the solar vertical circle can be used for direct determination of the conversion factor K, which changes very slowly with solar zenith distance. This method also has practical value.

6. Another method of eliminating higher-order scattering consists in determining three Stokes parameters: intensity, degree of polarization, and orientation of polarization, for example, relative to the sun's direction.

If higher-order scattering were completely absent, the plane of polarization would be directed precisely at the sun. On the other hand, with the cessation of twilight, the plane of polarization is oriented toward the effective layer of the primary segment at the very horizon on the solar vertical circle. The angle at a given point in the sky outside the solar vertical circle between the direction to the sun and the effective twilight segment β_0 can be calculated theoretically from

$$\iint B(z , A) f(\vartheta) \varphi(z , z_0)(P + P_0 \cos 2\alpha) \sin 2\beta d\sigma = 0 .$$

Angles β and α are shown in Fig. 5. For example, when the solar zenith distance is $100°$, $z_0 = 80°$, and $A_0 = 30°$, the angle β_0 has a rather high value ($33.0°$), and the point N, which represents the effective twilight segment, lies on the solar vertical circle at a height of $8°42'.7$.

It can be shown that, if we know from observations the three values I, P, and the angle of orientation γ, we can find the product I_1P_1, which applies only to primary twilight, from the sample expression

$$I_1P_1 = IP\,\frac{\sin\,(2\beta_0 - 2\gamma)}{\sin 2\beta_0}.$$

If we know the polarization components of the upper scattering indicatrix, we have P_1 for any angular distance from the sun and, therefore, the desired intensity I_1. Thus, higher-order scattering can be completely eliminated, and, if we know a number of other factors, which were mentioned above, we can immediately assess the optical properties of the upper atmospheric layers.

The analysis of twilight requires a comprehensive effort, using various instruments for simultaneous determination of the transmission coefficient and scattering indicatrix of the daytime sky, sounding of the lower atmospheric layers up to 40–50 km by artificial satellites, ozone determination, and, most of all, rapid and accurate determination of the three Stokes parameters.

The same apparatus can also be used to study the earth's dust cloud, which has still received very little attention, and also zodiacal light. All of this is considerably beyond the scope of the astronomical observatories and it cannot be undertaken by individuals. It is therefore very desirable to organize a zodiacal observatory with a small staff who could carry out the required work, which should be done at various locations and even in different hemispheres.

LITERATURE CITED

1.   V. G. Fesenkov, "On the earth's atmospheric shadow," Astron. zh., Vol. 26, No. 4 (1949).
2.   V. G. Fesenkov, Trudy Astrofiz. inst., Alma-Ata, Vol. 3, p. 214 (1962).
3.   V. G. Fesenkov, "On sounding the optical properties of the atmosphere with artificial satellites," Astron. zh., Vol. 43, No. 6 (1966).

ASTRONOMICAL AND ACTINOMETRIC VALUES OF STANDARD ATMOSPHERIC TRANSMITTANCE

K. S. Shifrin and G. L. Shubova

Analysis of data on the optical properties of the real cloudless atmosphere allowed us to propose as a standard of average conditions the Shifrin–Minin radiation model of the atmosphere [1]. With this model, we can calculate all of the necessary characteristics of the radiation field using the values of only three parameters: the optical thickness of the atmosphere τ_0 for a given wavelength, the meteorological visibility S_0 for the same wavelength, and the amount of precipitated water W. The radiation field is very sensitive to τ_0, which characterizes the aerosol turbidity of the atmosphere.

In this paper, we shall examine the statistical structure of the vertical atmospheric transmittance in the short-wave region, i.e., of the optical thickness τ_0 at $\lambda = 0.550\,\mu$.

Our problem was to determine the statistical characteristics of vertical transmittance directly from data on actinometric observations with filters.

Earlier [2-4], we examined the results of observations over Karadag in 1942, 1948, and 1949, and over Vladivostok, Irkutsk, Alma-Ata, and Kiev in 1935-1959. We have also studied the annual variation of transmittance over the above-mentioned locations [4].

In all of the measurements, the filters used cut out a region whose center was $\lambda = 0.577\mu$.

The main result of our earlier work is

$$\tau_{st} = 0.267; \quad \tau = 0.267; \quad \bar{\tau}_w = 0.246,$$

where τ_{st} is the transmittance for $\lambda = 0.577\,\mu$ in the standard radiation model [1]; τ is the arithmetic mean for all five locations; and $\bar{\tau}_w$ is the weighted mean, which takes into account the number of observations for each location. In all, there were 12,566 observations.

The standard value is greater than $\bar{\tau}_w$ and is the same as $\bar{\tau}$ (which, of course, was accidental). Should each of the values $\bar{\tau}_w$ and $\bar{\tau}$ be considered a true value? Unfortunately, this question cannot be answered at the present time. We must substantially increase the volume of initial data: the number of observations and especially the number of locations. These points must be selected to correctly reflect the different physicogeographical regions of the earth. Observations must be made over the oceans, which has not yet been possible for us. Nevertheless of the two figures obtained from our observations, $\bar{\tau}_w$ is the closer to the true value, if only because it gives considerably greater weight to Vladivostok, which reflects to a certain extent the optical conditions over the Pacific Ocean.

If $\overline{\tau}_w = 0.246$ is confirmed by more extensive and representative data, then τ for the standard model must be reduced from 0.267 to 0.246.

For $\lambda = 0.550\,\mu$, this means that the optical thickness must be converted from 0.300 to 0.277.

Optical thicknesses for $\lambda = 0.577\,\mu$ and transmission coefficients ($p = e^{-\tau}$) proposed by other researchers are [3]:

	p	τ		p	τ
Abbot	0.747	0.292	Lugin	0.724	0.323
Petty	0.785	0.242	Savost'yanova		
Lugin	0.795	0.229	(standards)	0.747	0.292
			Rabinovich	1.780	0.248

We see that these values agree approximately with the standard value 0.267.

But the values obtained by astronomers in photometric observations differ appreciably from the values obtained by actinometric observations.

Müller and Kempff [5] have obtained the most complete data on the transmission coefficient p. They cite 0.835 as the most probable value. This corresponds to $\tau = 0.180$. If we assume that we are dealing with radiation centered about $\lambda = 0.550\,\mu$, we can see that the optical thickness of the standard radiation model (0.300) is greater than the mean astronomical value by a factor of 1.67, and the mean actinometric value (0.277) reduced to $\lambda = 0.550\,\mu$ is greater by a factor of 1.54.

A probable cause of these differences is that astronomical observations are usually made on clear nights, while our measurements of direct solar radiation were made in the daytime, at times when the sun could be observed. It is known, for example, that (according to Rabinovich and Guseva [6] and Georgievskii [7]) transmittance has a clearly expressed daily variation with a maximum at night. To be sure, cases with comparatively small transmittance variations are also described in [7].

At the present time, it is difficult to explain with certainty the differences between the data of filtered actinometric and visual photometric measurements. But we consider the daily variation of transmittance and the selection of only clear nights for observations to be probable causes.

Standard Daytime Transmittance for Days with High Transmittance

Now let us analyze the most transparent states of the atmosphere. This will help us to shed some light on the problem.

Figures 1 and 2 show the probability densities for Vladivostok and Alma-Ata for all of the observation years (thin curves). Similar curves were obtained for the other locations.

The fact that the curves have several peaks means that we can isolate individual groups of states: high, intermediate, and low transmittance. Can the low value of standard τ used by astronomers be explained by the fact that they work under conditions of increased transmittance? To answer this, we made a special statistical study of the states corresponding to the first maximum on our curves for each location. In order to obtain a complete picture of the states with high transmittance, we were compelled, on the basis of the general nature of the curve, to construct with a certain degree of arbitrariness the right branch. These added parts are indicated by broken lines in Figs. 1 and 2.

The number of days N_1 with high transmittance and their percentage q of the total are given below. The number q can be considered a characteristic of the astroclimate of a given region.

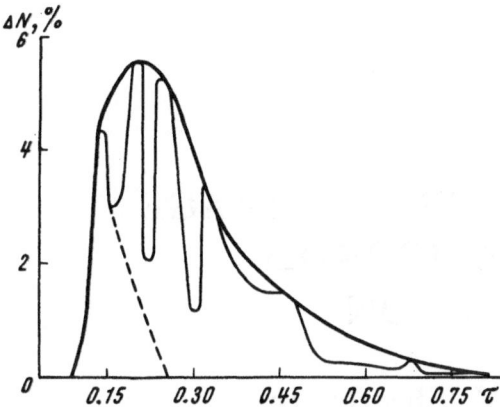

Fig. 1. Distribution of optical thickness for Alma-Ata, N = 1931, τ = 0.258, 1935-1950.

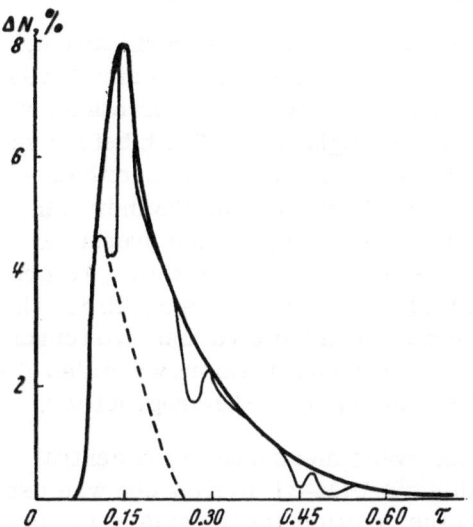

Fig. 2. Distribution of optical thickness for Vladivostok, N = 7234, τ = 0.232, 1938-1947.

	q, %	N_1
Alma-Ata	19.1	370
Vladivostok	26.5	1914
Irkutsk	11.4	42
Karadag	20.0	519
Kiev.	9.01	40

The total number of observations with high-transmittance states was 2885, or 23%.

We calculated the weighted mean optical thicknesses for these states for all of the observation points. The results are given below, where $\overline{\tau}^*$ is the arithmetic mean for all points and $\overline{\tau}_w^*$ is the weighted mean, taking into account the total number of observations at each point.

	$\overline{\tau}^*$
Alma-Ata	0.13
Vladivostok	0.15
Irkutsk	0.11
Karadag	0.18
Kiev.	0.20
$\overline{\tau}^*$	0.15
τ_w^*	0.15

If we convert to λ = 0.550 μ, we obtain $\overline{\tau}^*$ = $\overline{\tau}_w^*$ = 0.17. This is very close to the photometric value 0.18.

Thus, the mean standard optical thicknesses for daytime states with high transmittance agree with the photometric values. It must be emphasized that the states with high transmittance comprised only 23% of the total number of cloudless days.

LITERATURE CITED

1. K. S. Shifrin and I. N. Minin, "Toward a theory of nonhorizontal visibility," Trudy GGO, No. 68 (1957).
2. K. S. Shifrin and G. L. Shubova, "Statistical variability of vertical atmospheric transmittance," Izv. Akad. Nauk SSSR, sergiya geofiz., No. 2 (1964).
3. K. S. Shifrin and G. L. Shubova, "The variability of vertical transmittance," Trudy GGO, No. 170 (1965).
4. G. L. Shubova, "The annual variation of vertical atmospheric transmittance," Trudy GGO, No. 196 (1966).
5. Müller and Kempff, Photometrische Durchmusterung des nördlichen Himmels, Publ. Astron. Obs. Potsdam, Vols. 9, 13, 14, 16, 17.
6. Yu. I. Rabinovich and L. N. Guseva, "Experimental study of the spectral transmittance of the atmosphere," Trudy GGO, No. 118 (1961).
7. Yu. S. Georgievskii et al., The Intermediate Ray in the Atmosphere [in Russian], edited by G. V. Rozenberg, Izd. Akad. Nauk SSSR (1961).

RESULTS OF OBSERVATIONS OF STAR IMAGE
VIBRATION AT THE UZHGOROD
ASTRONOMICAL STATION

M. V. Bratiichuk and I. V. Shvalagin

Visual and photographic observations have been carried out for ten years at the station for observations of artificial earth satellites at Uzhgorod State University. In 1964, the university obtained an AVR-2 telescope, with which it began studies of atmospheric vibrations.

The observations whose results will be presented here were made photographically on the AVR-2, which has a 20-cm objective diameter and a 300-cm focal length. The star trails were photographed on 9 × 12 cm Agfa Spektral Gelb Rapid plates. Vibration observations were made on clear nights at three times: 1.5-2 h after sunset, at midnight, and 1.5-2 h before sunrise. At each time, 4-6 stars each with magnitudes of from 2.5 to 3.5 at various zenith distances from 0 to 85° at 20° intervals were selected. They were observed near the meridian at about the instant of upper culmination. The state of the nighttime sky, and the pressure, relative humidity, and the temperatures outside and inside were recorded each time. To prevent vibration of the telescope itself, clamps were attached 20 sec before photographing. The observations were made on windless nights. The telescope was mounted on a massive foundation rather far from the city and the highway, so that vibrations from ground tremors were also eliminated. Thus, it is hoped that the waviness of the trails was due to atmospheric heterogeneities.

The trails went over the entire length of the plates and were measured in the central part of the plate. A segment 3 cm long was selected for this. A UIM-21 microscope was used. The trail was oriented along the axis of the abscissas, and the positions of the axis of the ordinates was measured every 0.1 mm along the trail, i.e., at 300 positions. The ultimate aim of these measurements was to obtain the vibration amplitude as the standard deviation from the center line of the trail. The curvature of the trail was eliminated by the well-known method [1]. The measurement accuracy was estimated from three plates of excellent quality. The error in determining σ_z'' was found to be 0.06 arc sec.

In astronomical observations, it is desirable to have as many clear nights as possible. Data on observations from 1 April 1965 through 28 February 1966 are given below.

Total nights.................	334 or	100%
Completely clear.............	53	15.9
Completely cloudy............	204	61.1
Variable cloudiness		
(cloudy less than half of night)...	19	5.7
Variable cloudiness		
(cloudy more than half of night)..	58	17.3

176

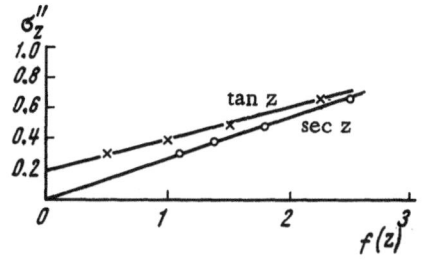

Fig. 1. Vibration amplitude versus sec z and tan z.

It should be emphasized, however, that these data are not typical for Uzhgorod, since it rained almost all summer in 1965.

At the present time, there are a number of papers on vibration amplitude as a function of zenith distance [2, 3]. Various authors have found various empirical relations to describe this. We decided to find the relation that agreed with our observations, and the following two turned out to most suitable:

$$\sigma_z'' = \sigma_0'' \sec z, \tag{1}$$

$$\sigma_z'' = \sigma_0'' + K \tan z, \tag{2}$$

where σ_z'' is the root-mean-square amplitude at zenith distance z and σ_0'' is the amplitude at the zenith.

We plotted 58 graphs for each relation. We found that 12 cases satisfied both relations, 33 cases only relation (2), five cases only relation (1), and eight cases did not satisfy either relation. An example where both relations were satisfied is shown in Fig. 1. Analysis of the results showed that the linearity of $\sigma_z''(\tan z)$ was disrupted on unstable nights. On the night of 21-22 October 1965, for example, the vibration amplitude at the zenith varied as follows:

Date	Universal time	Amplitude
21	18^h00^m	$0''.22$
21	22 00	0.40
22	03 00	0.26

The linearity of the relationship was preserved only in the evening (Fig. 2).

In [4], star-image vibrations were evaluated by σ_0'' values as follows:

Excellent.	$0.00—0''.20$
Good	0.21—0.30
Satisfactory	0.31—0.50
Poor.	over 0.50

For comparison, we divided our observations into similar groups.

Number of observation nights	46	or	100%
Excellent.	8		17
Good	21		46
Satisfactory	17		37
Poor.	0		

According to observation times,

Number of observation times	76	or	100%
Excellent.	18		23
Good.	34		45
Satisfactory	22		29
Poor	2		3

The dependence of the root-mean-square vibration amplitude upon tan z was found as follows. All of the σ_z'' values were plotted against tan z (Fig. 3). As can be seen from Fig. 3,

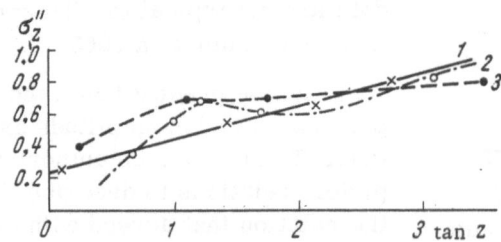

Fig. 2. Vibration amplitude versus tan z for
three observation times: 1) evening; 2) morning; 3) midnight.

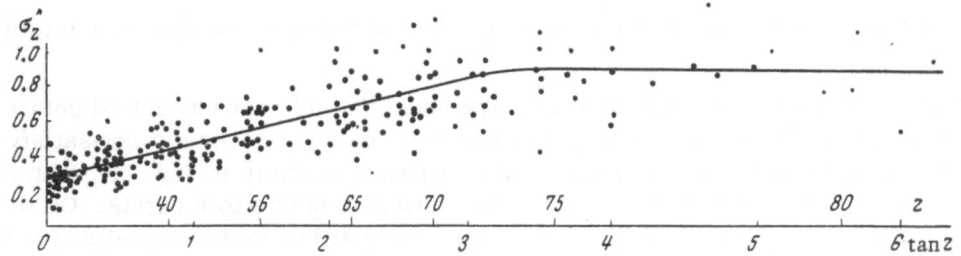

Fig. 3. Root-mean-square vibration amplitude versus tan z for all periods
of observation.

this dependence is linear up to $z = 72°$, after which the line is less steep. Satisfactory vibration
amplitude is observed up to zenith distance $z = 55°$. The mean values as calculated graphically
for the entire observation period are:

$$\sigma_0 = 0.28; \quad \sigma_{40} = 0.40; \quad \sigma_{65} = 0.62.$$

The angular coefficients K varied from $K_{min} = 0.02$ to $K_{max} = 0.60$, i.e., within rather wide
limits. The most probable K for our observations was 0.12.

We found that the vibration amplitude at the zenith was lower at high temperatures than
at low. No dependence of σ_0'' upon pressure or relative humidity was observed.

It is well known that when atmospheric heterogeneities smaller than the objective diameter pass in front of the telescope, the star image can be shifted at the focus of the telescope.
They impair the quality of the diffraction pattern. By photographing and measuring star-
image trails, we can take into account heterogeneities that are larger than the objective diameter. It should be emphasized, however, that we cannot find all of these shifts by measurement. Elementary calculations, taking into account the scale of our photographs and diffraction, show that if deviations occur more frequently than every 0.1 sec, they cannot be determined by this method. In other words, the high-frequency part of vibration widens the trails
and remains undetermined, since the trail width is still a function of star magnitude. In view
of this, we attempted to take into account the high-frequency component by photographing the
trails of Echo-type satellites.

LITERATURE CITED

1.   Yu. Yu. Beruchka, Izv. GAO, Pulkovo, Vol. 21, Issue 6, No. 165 (1960).
2.   I. G. Kolchinskii, Astron. zh., Vol. 34, No. 4 (1957).
3.   O. B. Vasil'ev, in: Optical Instability of the Earth's Atmosphere [in Russian], Nauka (1965).
4.   O. P. Vasil'yanovskaya, in: Optical Instability of the Earth's Atmosphere [in Russian],
      Nauka (1965).